T0138482

Analyzing Animal Societies

Analyzing Animal Societies

Quantitative Methods for Vertebrate Social Analysis

Hal Whitehead

The University of Chicago Press :: Chicago and London

Hal Whitehead is the University Research Professor in the Department of Biology at Dalhousie University in Halifax, Nova Scotia. He is author of *Sperm Whales: Social Evolution in the Ocean* (2003) and coeditor of *Cetacean Societies: Field Studies on Dolphins and Whales* (1999), both published by the University of Chicago Press.

The University of Chicago Press, Chicago 60637
The University of Chicago Press, Ltd., London
© 2008 by The University of Chicago
All rights reserved. Published 2008
Printed in the United States of America

17 16 15 14 13 12 11 10 09 08   1 2 3 4 5

ISBN-13: 978-0-226-89521-5 (cloth)
ISBN-13: 978-0-226-89523-9 (paper)
ISBN-10: 0-226-89521-1 (cloth)
ISBN-10: 0-226-89523-8 (paper)

Library of Congress Cataloging-in-Publication Data

Whitehead, Hal.
  Analyzing animal societies : quantitative methods for vertebrate social analysis / Hal Whitehead.
      p. cm.
  Includes bibliographical references.
  ISBN-13: 978-0-226-89521-5 (cloth : alk. paper)
  ISBN-10: 0-226-89521-1 (cloth : alk. paper)
  ISBN-13: 978-0-226-89523-9 (pbk. : alk. paper)
  ISBN-10: 0-226-89523-8 (pbk. : alk. paper) 1. Vertebrates—Behavior—Mathematical models. 2. Animal societies—Mathematical models. 3. Social behavior in animals—Mathematical models. I. Title.
  QL751.65.M3W48 2008
  591.7'82—dc22
                                             2007042457

♾ The paper used in this publication meets the minimum requirements of the American National Standard for Information Sciences—Permanence of Paper for Printed Library Materials, ANSI Z39.48-1992.

To my parents, Denis and Frankie Whitehead

Contents

Acknowledgments

This book is the result of the ideas and energy of many scientists who have pondered the social relationships of animals. The conceptual framework of Robert Hinde has been particularly influential and helps to structure the book. David Lusseau introduced me to a stream of papers and ideas, especially on network analysis. His creativity has had a major impact on this book. We worked together closely, trying to make network analysis effective in the description of social systems (Section 5.3) and in community delineation (Section 5.7). Like David Lusseau, Lars Bejder came to my laboratory from Otago University bringing new ideas and techniques, especially on permutation tests and measures of social complexity. Ideas about measuring the influence of culture in animals (Section 7.6) were developed during collaborations with two other Dalhousie colleagues, Bob Latta and Luke Rendell.

I am grateful to Shane Gero, Meaghan Jankowski, and two anonymous reviewers for their constructive reviews of the book proposal. The book was greatly improved by the comments of two anonymous reviewers and John Fryxell, as well as reviews of Chapters 1 to 3 by Darren Croft, Shane Gero, Shannon Gowans, and David Lusseau and of Chapter 5 by David Lusseau. Shannon Gowans suggested the inclusion of Section 1.2 and provided an outline for that section, which I have largely followed. Andrew Horn kindly advised on, and gave many useful suggestions for, Section 1.4. Section 5.4 on dominance hierarchies was reviewed thoroughly by Han de Vries, whose authoritative comments were particularly helpful. Katie McAuliffe advised on field methods for terrestrial mammals.

For permission to describe unpublished ideas or techniques, I am very grateful to Colin Garroway (subdivision of fission–fusion societies, Section 6.2), David Lusseau (several network analysis techniques, Sections 5.3 and 5.7), and Richard Wrangham (definition of bond, Section 4.10). Maarten Vonhof gave permission to use his data on bats. Peter Chen of Noldus provided an inspection copy of MatMan.

Thanks to Shane Gero for Fig. 4.7 and David Lusseau for Fig. 5.17.

I am particularly grateful to Emese Kazár for her illustrations of social animals. These link the products of the methods in this book back to their root. I also thank Paul Bentzen, Brock Fenton, Ian McLaren, Susan Perry, and Marten Vonhof for checking Emese's illustrations of their study animals.

Christie Henry of the University of Chicago Press has been consistently encouraging throughout. Thanks to Christie, Tisse Takagi, and the copyediting team for dealing with my errors, and efficiently making a book from my manuscript.

Conventions and Abbreviations

Important terms are defined in the glossary (Appendix 9.1) and are italicized when first discussed in detail.

Important mathematical equations are given up to the level of the multiple summation ($\sum\sum$). For more complex mathematical details, see cited references.

Some more specialized techniques and mathematical details are given in appendices.

Unless taken directly from another publication, identifiers of individual animals are given in upper case (e.g., "B", "JOE") or preceded by the number sign if numeric (e.g., "#305"), and are not italicized.

Scientific names of species are given once in each section, either in the first table or figure captions or, if there is no table or figure, when first mentioned in the main text. Given the current uncertainty in delphinid taxonomy and nomenclature, and following current practice, I give only the generic name for *Tursiops* spp. even when species names were provided in the papers being referenced.

AIC	Akaike information criterion
ANOVA	analysis of variance
CCC	cophenetic correlation coefficient
CV	coefficient of variation
ID	individual identification
MANOVA	multivariate analysis of variance

MDSCAL	nonmetric multidimensional scaling
P	probability of obtaining the true value of the test statistic, or a more extreme value, under the null hypothesis
PIT	passive integrated transponder tags
r	correlation coefficient
SD	standard deviation
SE	standard error
Var	Variance
$\bar{x}$	mean
$\sum$	summation over
τ	time lag

1   Analyzing Social Structure

1.1 Introduction

Social structure synthesizes a vital class of ethological and ecological relationships—those among members of the same species whose ranges overlap. This book is about how to study, analyze, and model social structure (here synonymous with social system, social organization, and society). It is intended to assist biologists studying social structures.

Social structure is founded on behavioral interactions among individuals. Without identification of individuals, analyses of social structure are constrained to be simplistic. Studies of social behavior in vertebrates usually identify individuals, whereas those of invertebrates usually do not. Hence "vertebrate" in the book title—I almost always assume that data refer to individually identified animals. However, the methods I describe could be effectively applied to invertebrates in many cases, and I will consider a few techniques that do not require individual identification.

This is a book of methods intended for the scientist or student who can identify members of an animal population and can record their interactions or associations. There are several fine guides to appropriate methods of social observation (e.g., Altmann 1974; Martin & Bateson

2007 Lehner 1998). Following this advice, social interactions can be observed systematically and without much bias, but then what? How do we go from records of "ID#302 groomed/grouped with ID#127" to a model of social structure? How do we test a hypothesis, such as that grooming is reciprocal, based on theory or discoveries on other populations? There has been remarkably little guidance on the analytical side of social analysis. Fine data sets are collected, but frequently analyses are not even attempted, and when they are, they are often suboptimal.

Following a brief consideration of technical, principally statistical, matters (Chapter 2) and data collection (Chapter 3), the book principally focuses on the production of valid, quantitative models of social structure from such data. I use as a backbone Hinde's (1976) conceptual framework for the study of social structure (Section 1.6). This is based on interactions between individuals that are integrated and abstracted to describe relationships between members of dyads, where a dyad is a pair of individuals (Chapter 4) and the relationships among all dyads in the population form its social structure (Chapter 5). These two chapters are the heart of the book. In some cases, the division of the material between these two chapters is rather arbitrary. For instance, plots of relative measures of relationship (e.g., Fig. 4.7) indicate both the nature of the relationships and the form of social structure. I then consider whether and how we can compare different societies (Chapter 6), before ending in a less comprehensive vein by discussing methods of examining the evolutionary and ecological links between social structures and other biological attributes (Chapter 7) and how social analysis may develop in the future (Chapter 8). The appendices include definitions of important terms used in the book (Appendix 9.1), books that I have found useful in carrying out social analyses (Appendix 9.2), some information on how computer packages can be used for social analyses (Appendix 9.3), and a few statistical derivations (Appendixes 9.4 and 9.5). To provide larger-scale guidance, I have provided general recommendations in text boxes.

An important decision in writing this book has concerned the level of statistical detail to include. Statistical methods used in social analysis range from the trivial (e.g., when constructing sociograms) to complex modern techniques such as Markov chain Monte Carlo methods. Aiming at what I perceive to be the median level of statistical expertise of those who study animal societies, I include within the main text of the book descriptions of the techniques, how they are used, when they are appropriate, when they are not appropriate, output options, and potential difficulties. Where derivation or calculation of a measure is fairly straightforward, I give formulas (roughly up to the level of multiple summations, $\Sigma\Sigma$). More-detailed statistical matters and more-complex procedures are covered in the technical works cited and in appendices.

Nearly all of the available techniques of social analysis require computer programs. Wherever such techniques are available, I show in Appendix 9.3 how reasonably easy-to-use computer programs can be used to carry out the analyses.

This book is *not* a statistical treatise; a guide to field or laboratory methods of studying the social behavior of animals; a collection of computer programs; a synthesis of results on vertebrate social structures; or a review of vertebrate social evolution. All of these are or would be most valuable, but this is a book about methods of social analysis. It concentrates on the most fundamental methods. There are many examples in the literature of quite complex and creative analyses of vertebrate social structures. There are too many to cover well, however, and most are highly situation dependent, so I stay with those techniques that I think will have reasonably broad usage.

The methods are illustrated by examples from a variety of taxa. I have often chosen parts of the available data in ways to best illustrate a particular analytical method. This means that different parts of the same data set are used at different places in the book. It also means that, while the results presented in the examples generally represent the social systems from which the data were collected, those interested in these social systems should use the original publications, which I cite, not the examples in this book.

I have tried to keep the book relatively brief. I hope that those who need guidance and information on methods of social analysis can find much of what they need.

1.2 *What Is Social Structure?*

Later in this chapter (Section 1.6), I introduce a formal conceptual framework of social analysis, but, principally for the benefit of those new to social analysis, I provide here a brief summary of the object and goal of this field. Social structure is a synthesis of how individuals interact with each other. Ideally, a description of the social structure of a population captures the nuances of individual differences in social behavior but also efficiently summarizes the actions of the individuals and their relationships with each other. A good description of social structure should be sufficiently general to permit comparisons with social structures of other populations and correlations between social structure and nonsocial factors such as population density and predator pressure.

1.3 Why Social Structure Is Important

For all but the most solitary asexual organisms, nearby conspecifics are vital elements of the environment. Individuals of the same species often compete with each other for resources. In some cases, they may use each other as resources, for instance, in cannibalism. Alternatively, they may cooperate in attaining resources or defending either resources against conspecifics or themselves against predators. They may mate with one another or care for one another. Thus, social structure is often a key determinant of population biology, influencing fitness, gene flows, and spatial pattern and scale (Wilson 1975). Through these routes, social structure often becomes an important element in the management and conservation of a species (Sutherland 1998). For instance, poaching affects the health of African elephant (*Loxodonta africana*) populations not only through the number of animals killed, but also because in this highly social species the reproductive success of individuals depends on social relationships with others (Poole & Thomsen 1989).

With patterns of mortality, reproductive success, and dispersal strongly affected by social structure, one expects that not only will social structures evolve into forms adaptive for their members, but also that these forms will influence the evolution of social and other traits. For instance, interactions with relatives may affect inclusive fitness and thus drive social evolution (Hamilton 1964), and the evolution of cooperation among nonrelatives also depends on the form of social structure (Trivers 1985). It is believed that social structures shaped the evolution of sexual size dimorphism (e.g., Lindenfors et al. 2002), signaling systems (Bradbury & Vehrencamp 1998), and cognition (Byrne & Whiten 1988).

The rate and pattern of spread through a population of anything that is transmitted by individual-to-individual proximity or interaction will depend on social structure. Thus, network analyses of social systems are used to study information flow (e.g., Lusseau 2003) and disease transmission (e.g., Rogers et al. 1998).

So social structure is important, and scientists should study it.

1.4 Conceptualizing Animal Societies: A Brief History

The study of the structure of vertebrate societies has a diffuse history. Scientists studying particular species found models to guide their analysis from a wide variety of sources or invented their own. In this section, I sketch the major paths of social analysis. Crook (1970) provides a nice summary of the history of social analysis until 1970, and Roney and

DES

SOCIÉTÉS ANIMALES

ÉTUDE DE PSYCHOLOGIE COMPARÉE

PAR

ALFRED ESPINAS

ANCIEN ÉLÈVE DE L'ÉCOLE NORMALE, AGRÉGÉ DE PHILOSOPHIE
PROFESSEUR DE PHILOSOPHIE AU LYCÉE DE DIJON

FIGURE 1.1 Frontispiece of Espinas' (1878) treatise on an animal societies, from a 1924 reprint.

Maestripieri (2003) discuss more recent developments in the study of primate societies.

Although there were earlier perspectives on animal societies, a landmark, at least in hindsight, was the publication of *Des sociétés animales* by Espinas (1878) (Fig. 1.1). Espinas provided a history of social theory and summarized what was known of animal social life, categorizing animal societies. His approach, which used evolutionary and ecological ideas as well as perspectives from sociology, looks quite modern in retrospect. He noted that animal societies arose through the "habitual reciprocity" of the actions of its members, and that they seemed adapted to local conditions but were not clearly related to phylogeny. The ideas of Espinas, although influential in the development of sociology, were largely ignored by zoologists, as were those of some early twentieth-century scientists who also examined animal societies from both sociological and evolutionary perspectives (Crook 1970).

Instead, in the mid-twentieth century, an ethological outlook on social structure based on the motivational and physiological causation of behavior became dominant. The perspective is clear from the title of Lorenz's classic paper "Der Kumpan in der Umwelt des Vogels (Der Artgenosse als auslösendes Moment sozialer Verhaltungsweisen)," or "The companion in the bird's world: the fellow-member of the species as a releasing factor of social behavior" (Lorenz 1935, summarized in English in Lorenz 1937). Individuals, through their presence or behavior, were seen as releasers of each others' instinctive behavior. Nonhuman social structure was seen as the sum of these instinctive reactions to the "Kumpan" (roughly translated as "companion"). It lacked the "deeper and nobler bonds" of human society.

This early ethological view of vertebrate social behavior "based on the releaser system" (Tinbergen 1953) was in some respects in competition with other paradigms, including the ideas of the even more mechanistic U.S. behaviorist school. In the 1950s other ethologists began seriously to question Lorenz's motivational approach (Crook 1970). A range of perspectives was appearing. For instance, Thompson (1958) considered social behavior from an evolutionary viewpoint, emphasized communication and interaction as the basis of social structure, and examined the attributes of animal groups, which might include bonds and altruism. The paradigm was shifting, partly due to an increasing attention to primates in place of Lorenz's birds. With a focus on primates, it is more natural to make comparisons with humans, so ideas were imported from the human social sciences. Some anthropologists, sociologists, and psychologists began examining primate social structures, hoping to learn about the roots of human society.

In the 1960s and early 1970s, role theory, developed in social psychology and anthropology, became an important paradigm in primatology (Roney & Maestripieri 2003). From this perspective, individuals have distinctive roles in society, and social structure is the outcome of behavior resulting from these roles. In the 1970s and early 1980s, primatologists imported another set of models, this time from sociology. Sociometric analyses, and the related block-model approach, focused on dyadic relationships rather than individuals (Roney & Maestripieri 2003). In some respects, this completed a loop because fundamental concepts of sociology were developed from models of nonhuman societies.

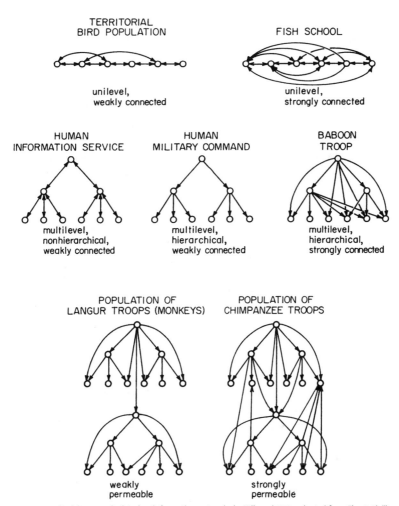

FIGURE 1.2 Social groups depicted as information networks by Wilson (1975; adapted from Fig. 2-3) "in order to illustrate variation in several of the qualities of sociality."

For instance, Emile Durkheim, one of the fathers of sociology, was strongly influenced by Espinas (1878).

In his monumental book *Sociobiology*, Wilson (1975, pp. 16–19) (Fig. 1.2) drew together many of these threads, listing 10 "qualities of sociality": group size, demographic distributions, cohesiveness, amount and pattern of connectedness in communication, permeability or movement between social groups, compartmentalization or the degree to which the population contains distinct social units, differentiation of roles, integration of behavior, information flow, and fraction of time devoted to social behavior. All of these are important qualities of societies and can be used in comparative studies or in attempts to classify societies (Section 1.8).

Social structures have not escaped the drive to classify. For instance, Espinas (1878) produced a simple classification based on whether associations among individuals are active or passive and are colonial or free-living and the function of aggregation, and others followed with systems that were more or less divisive (Wilson 1975, pp. 16–19). Such blanket classifications of animal social systems were unsatisfactory in several respects. Divisions were at least partially arbitrary. If they were of fine scale, it was often unclear into which category the social structure of a given population of animals should be placed, and populations could appear to switch categories over time. With broad categories, patently dissimilar social systems might end up in the same bin. Categorization has been more successful on subsets of animal social systems, such as those of the social insects (Michener 1969) or primates (Crook & Gartlan 1966; Kappeler & van Schaik 2002) or the mating systems of mammals (Clutton-Brock 1989). In each of these cases, there are apparent categories with quite clear discontinuities between them (Section 1.7 and Chapter 6).

Whereas the sociobiologist Wilson (1975) discussed the "top-down" classification and measurement of social structure, ethologists took a more "bottom-up" approach. Having moved beyond Lorenz's releasers, they melded their descriptive methods to ideas imported from sociology. Hinde's (1976) conceptual framework for the study of social structure (Section 1.6), rooted in dyadic interactions, cemented the position of the sociometric approach as fundamental to studies of primate social structure, and it has a central role in structuring this book.

Initially at least, this conceptually rich approach was principally applied to primates. Studies of social structure in other vertebrates went little beyond Wilson's (1975) first two qualities of sociality: group size and demography. Although primatologists were using methods developed for humans, there was a deep reluctance among some scientists to infer complex sociality among other vertebrates. Although it was considered likely, and a useful research hypothesis, that individual primates recognize and form distinct relationships with other individuals, distinctive relationships among individuals of other species were rarely considered. Individual recognition and dyadic relationships may not be important elements of the social structure of most pelagic fishes. There are other species, however, whose social complexity is not very different from that of primates. The gulf is illustrated by perspectives from about 1980 on the social ecology of two vertebrate orders that are now recognized as having quite similarly high levels of social complexity (Connor et al. 1998). On one hand, Nagel (1979, p. 316) asserted "a primate group can certainly be described as a complex of interacting members," whereas Gaskin (1982, p. 151) concluded "statements that assume the existence of a high order of social evolution in the Cetacea are, frankly, not really supported to any extent."

In the 1980s and 1990s, the gulf between the productive course of primate social analysis and the depauperate studies of the social structure of most other vertebrates began to narrow. This was due to a largely one-way flow of personnel and ideas. Scientists trained as primatologists or by primatologists began detailed and highly illuminating studies of other species (e.g., Clutton-Brock et al. 1982; Connor et al. 1992), and important ideas from primatology, such as Hinde's framework (Section 1.6), began to influence nonprimate social analysis (e.g., Le Pendu et al. 1995). Meanwhile, other ideas, such as the game-theoretic concept of evolutionarily stable strategies developed in economics, were placing the social behavior of both primates and nonprimates in new and productive perspectives (e.g., Trivers 1985).

Despite their slow start, toward the end of the twentieth century, nonprimatologists became active in developing new quantitative methods of social analysis. This is perhaps partly because they were less constrained by "standard practices" and partly because nonprimate study populations are often larger, making quantitative methods more valuable. For instance the temporal methods described in this book (Sections 4.6 and 5.5) had their roots in Myers' (1983) analysis of shorebird society and Underwood's (1981) work on ungulates. Consequently, during the 1980s and 1990s, whereas studies of primate societies developed into new conceptual directions (including culture, politics, and areas of experimental psychology), those of nonprimate vertebrates became generally more quantitatively sophisticated (Whitehead & Dufault 1999).

An important approach to the analysis of vertebrate social systems—network analysis—has recently been introduced. Used for some time in sociology and other scientific disciplines, with very few exceptions (e.g., Maryanski 1987), network analysis was not applied to nonhuman vertebrate societies until 2003. Then, network analysis caught the attention of scientists studying dolphins (Lusseau 2003) and fish (Croft et al. 2004), with applications to primates (Flack et al. 2006) following.

In this book, I describe methods of vertebrate social analysis, often developed on nonprimate species, from a conceptual perspective that stems largely from studies of apes and monkeys.

1.5 Ethology and Behavioral Ecology

Studies of animal social structure are generally approached from one of two paradigms—*ethology* or *behavioral ecology* (which is related to, and sometimes synonymous with, sociobiology). Ethologists take what might be called a bottom-up approach to animal behavior, "attempting to start their analyses from a secure base of description" (Hinde 1982, p. 19), and then examining immediate causation, development, function, and/or evolution—Tinbergen's (1963) four "whys." Ethology has strong links to a number of areas of psychology (Hinde 1982).

Behavioral ecology is a subdiscipline of Darwinian evolutionary biology, being concerned principally with the function of behavior and how a particular behavioral pattern influences survival and reproductive success (Krebs & Davies 1991). Thus, an ethologist studying social structure may try to work out how patterns of interactions among individuals are organized and change with time, whereas a behavioral ecologist will measure fitness differences of females in different-sized groups or between those males who defend territory and those who do not. The two approaches are linked, and I draw from both.

With simple or easily classifiable social traits (such as some mating systems), the behavioral ecological approach can proceed from relatively straightforward field or laboratory data, with the primary technical challenge being the measurement of fitness. However, the social structures of many vertebrates are not easily described or classified. Addressing functional questions using poorly supported social measures such as ill-defined "group sizes" of different species can be misleading (e.g., Connor et al. 1998). In such situations, the ethological approach—accurately describing and modeling social structure—should precede behavioral ecological analysis. Thus, my approach to social analysis is primarily based on that of the ethologists.

By using the concept of the graphical network (Newman 2003b), however, the ethological and behavioral ecological perspectives can be integrated. Individuals are represented by nodes or points on a graphical representation (e.g., Fig. 5.5). The nodes are linked in ways that describe their interactions and relationships. The primary work of the ethologist, then, is in producing this representation and then analyzing and describing the network (Section 5.3). The behavioral ecologist asks "why"? Why are nodes linked in the way that they are? Why does the network take the overall form that it does? The behavioral ecologist asks ecological "whys": Given an individual's genes and environment—its physical, ecological and social environment—why does an animal

Table 1.1 Definitions of Social Structure and Related Terms from the Literature

Social structure	"The composition of groups and the spatial distribution of individuals" (Rowell 1972, 1979)
	"Pattern of social interactions and the resulting relationships among members of a society" (Kappeler & van Schaik 2002)
Social organization	"The pattern of interactions between individuals, a description of behavior" (Rowell 1972, 1979)
	"Size, sexual composition and spatiotemporal cohesion of society" (Kappeler & van Schaik 2002)
	"Union of overlapping social niches" (Flack et al. 2006; supplementary information)
Social system	"Set of conspecific animals that interact regularly and more so with each other than with members of other such societies" (Kappeler & van Schaik 2002; considered synonymous to "society"; roughly equivalent to "community" in the terminology of this book)
Society	"A group of individuals belonging to the same species and organized in a cooperative manner" (Wilson 1975, p. 595)
Surface structure	"Nature, quality and patterning of relationships" [a] (Hinde 1976)
Structure	Generalizations of surface structure across populations (Hinde 1976)
Deep structure	Dynamic understanding of patterns of relationships (Hinde 1976)

[a]This is the definition used in this book.

behave as it does? But she also asks evolutionary questions: How do the physical, ecological, and social environments interact with behavior to change the distribution of genes within a population? To consider the behavioral ecologist's questions, we need an accurate representation of the social network or at least an understanding of the limitations of the representation. Conversely, with her analysis complete and a representation of a social network in front of her, any thinking ethologist will ask: Why?

1.6 Hinde's Ethological Conceptual Framework of Social Structure

There are a number of concepts and definitions of social structure in the ethological and behavioral ecological literatures (Table 1.1). Most are sensible and reasonable, but there is one that has proved to be the most logically consistent and empirically useful. It is also the one that has probably been the most frequently cited.

Robert Hinde (1976) produced his conceptual framework for the analysis of animal societies from an ethological perspective based on his observations of captive primates as well as a thorough knowledge of the methods used to study human societies. Hinde's framework is applicable for very rich studies of complex societies (such as humans or chimpanzees, *Pan troglodytes*) as well as much more basic studies of species with simple societies (such as some rodents) or the limited data sets that are available for more cryptic species (including arboreal and

Social Structure
nature, quality, and patterning of relationships

⇑

Relationships between Individuals
content, quality and patterning of interactions

⇑

Interactions among Individuals

FIGURE 1.3 Summary of Hinde's (1976) framework for the analysis of animal societies.

aquatic animals). Particularly in the case of primates, Hinde's framework has formed the conceptual basis of a wide range of studies of social structure (Roney & Maestripieri 2003).

Hinde's framework is illustrated in Figs 1.3 and 1.4. Figure 1.3 summarizes the essence of Hinde's framework, and Fig. 1.4, taken directly from Hinde (1976), introduces some of the major factors that may influence social structure and the abstractions that may be involved in studying it.

At the fundamental level is the *interaction*: When the presence or behavior of one individual is directed toward another or affects the behavior of another—in Hinde's words, "what the animals are doing together (its content) and how they do it (quality)." Although behavioral interactions are usually among members of the same species, this is not always the case, for instance, in mixed-species bird flocks. A *relationship* between two individuals comprises the content, quality, and patterning of their interactions, in which patterning is with respect both to each others' behavior and to time. Finally, the *social structure* of a population comprises the nature, quality, and patterning of the relationships among its members.

Hinde's framework does not include any causal direction: Social structure is the result of interactions, but type, quality, and patterning of interactions may be influenced, sometimes strongly, by social structure (Hinde 1976; Fig. 1.3). Some of the other factors that may affect social structure at different levels of the framework are shown around the periphery of Fig. 1.4. From the perspective of this book some of the

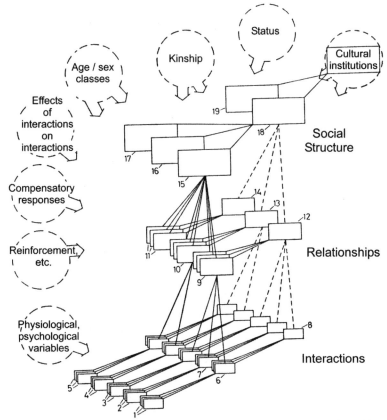

FIGURE 1.4 Hinde's (1976) framework for the analysis of animal societies. Interactions, relationships, and social structures are shown as rectangles on each level, with successive stages of abstraction moving from left to right. The circles note independent or intervening factors that may operate at different levels. Cultural institutions, having a dual role, are shown in both a circle and a rectangle. For nonhuman primates, these numerically coded examples might represent the following:

1. Instances of grooming between a mother A and her infant B.
2. Instances of nursing interactions between A and B.
3. Instances of play between A and B.
4. Instances of grooming between female A and male C.
5. Instances of copulation between A and C.
6. First-stage abstraction: schematic grooming interactions between A and B. Abstractions of grooming interactions between other mother–infant pairs are shown behind, but the specific instances from which they were abstracted are not shown.
7. First-stage abstraction: schematized nursing interactions between A and B. Abstractions of nursing interactions of other mother–infant pairs are shown behind.
8. Second-stage abstraction: schematized grooming interactions between all mother–infant pairs in the population.
9. Mother–infant relationship between A and B. Mother–infant relationships of other mother–infant pairs are shown behind (but connections to grooming, nursing, etc., interactions are not shown).
10. Consort relationship between A and C. Other consort relationships are shown behind.
11. Specific relationship of another type (e.g., peer-peer).
12, 13, 14. Abstraction of mother–infant, consort, and peer–peer relationships. These may depend on abstractions of the contributing interactions.
15. Surface structure of the troop containing A, B, C, etc.
16, 17. Surface structure of other troops (contributing relationships not shown).
18. Abstraction of structure of set of troops containing A, B, C, etc. This may depend on abstractions of mother–infant, etc., relationships.
19. Abstraction of structure of a different set of troops (from another environment, species, etc.).

(From Hinde 1976 with a few changes in terminology, and from which most of this caption is taken.)

more significant are age/sex classes, kinship, and cultural institutions (Chapter 7). These may help us to "explain" the patterns found at each level (Hinde 1976); for instance, kinship and the age/sex class of the individuals may explain much of the variance in patterns of interactions or relationships, and aspects of culture, such as conformism, can have a strong influence on social structure.

Using Hinde's framework, the study of social structure is based on identifying individual animals and recording their interactions, an ethological bottom-up approach. The analytical challenge is to find and synthesize measures of content, quality, and patterning at the levels of both interaction and relationship.

1.6.1: Changes from Hinde's Terminology. Hinde (1976) calls the third level of his framework "surface structure": "that which is apparent in empirical data." Thus "surface structure" is in the eye of the beholder—literally the data of the scientist—who hopes that it approximates the true social structure of the population. The distinction between the true nature of a parameter, entity, or system and its estimated nature is always important, and particularly so in the case of complex systems that may be hard to model or conceptualize. Following usual practice, however, throughout this book I will use the same name for the true value or nature of a parameter, entity, or system and its estimated value or nature. Thus "social structure" will be used in place of "surface structure." Sometimes I use the adjectives "true" or "estimated" to emphasize the difference. Readers should be aware, however, that the true and estimated will always differ, often substantially and especially so in the case of complex systems such as social systems. Hinde (1976) defines and discusses two other forms of social structure: (1) "structure"—generalizations from studies of social structure in several populations; and (2) "deep structure"—the principles of societal organization (Table 1.1). In this book, I consider both generalizations and principles, but, aiming for conceptual simplicity, refer to them as such, leaving "social structure" for the actual or inferred society of a particular population.

Another change from Hinde's terminology is the replacement of "group" by "population." "Group," as used by primatologists, including Hinde (1976), refers to a set of individuals that is largely behaviorally distinct from all others, but when referring to most other vertebrates, "group" has a very different, and much more transitory, connotation (Whitehead & Dufault 1999). I will call a largely behaviorally closed set of animals in which most individuals interact with most others (the primate "group") a *community*. For some species (including some primates, such as orangutans, *Pongo pygmaeus* [Mitani et al. 1991]), there are no clear entities corresponding to the community. For the purposes of this book, I will define *population* as a set of animals such that the great majority of interactions that involve at least one member of the population involve only members of the population. Thus, a population could consist of one or more communities or a spatially discrete set of individuals. The key feature is that within-population interactions greatly outnumber those between members of different populations.

As another attempt to simplify terminology, I will treat *social organization*, *social system*, and *society* as synonyms of social structure. Some authors (e.g., Rowell 1972, 1979; Kappeler & van Schaik 2002) have given some of these terms distinct definitions to clarify aspects of social analysis (Table 1.1). These distinctions can be useful, but on balance I think simplicity is to be preferred. In most published studies on vertebrates, social structure, social structure, and society are neither defined nor distinguished.

1.7 Other Definitions and Concepts of Social Structure

There are other ethologically oriented definitions of social structure, such as those of Rowell (1972, 1979) and Kappeler and van Schaik (2002), that, although they differ in specifics from that of Hinde, have a conceptually similar ethological orientation toward animal societies (Table 1.1). They are generally more complex than Hinde's framework and have been less widely used.

At least two other concepts of animal social structure differ from that used by Hinde, Rowell, Kappeler and van Schaik, and this book, and appear in the literature on animal societies. They are principally found in work rooted in the behavioral-ecological perspective on animal behavior. Studies of invertebrates, particularly social insects, tend to focus on emergent properties, such as divisions of labor (including "castes"), and patterns of alloparental care and aggregation (e.g., Michener 1969). Species may then be allocated to categories of social structure such as "communal" or "eusocial" with the hope of identifying drivers of these patterns, such as ecological attributes or kinship. From this perspective, a primary challenge is seen as the classification of social systems, given reasonably easily obtained knowledge on the emergent properties (Costa & Fitzgerald 1996).

Some species, including many pinnipeds and amphibians, principally congregate to breed. Most studies of the social behavior of these animals are of breeding behavior, and mating systems are used to classify their social structure. As with the studies of social structure in insects, ecological and other factors, such as kinship, are considered as potential drivers of the different types of mating systems (e.g., Emlen & Oring 1977).

These approaches to the study of social structure are appropriate with such species, but most vertebrate social structures cannot easily be classified by large-scale properties and include much more than breeding behavior. Thus, I generally advocate an initial use of the bottom-up, ethological approach to studying social structure that was conceptualized by Hinde. The results of such analyses can be linked to data on mating systems or other attributes of populations (e.g., Kappeler & van Schaik 2002).

More recently, another approach has emerged, the *social niche*. Rather than focusing on dyadic relationships as in Hinde's (1976) framework, the social niche is an individual characteristic: the "vector of behavioral connections in the set of overlapping social networks in which it participates" (Flack et al. 2006). Social niches have similarities with ecological niches. They vary in quality, both between individuals and with time, affect reproductive success, and interact with one another. Individuals may be able to "construct" social niches as they can ecological niches (Laland et al. 2000), potentially improving reproductive success. The statistics of network analysis (Section 5.3) provide useful quantitative descriptors of the social niche. Social organization is then the union of overlapping social niches (Flack et al. 2006). The concept of the social niche is sufficiently new that its true significance is unclear. I believe that it has considerable potential, however, and may, at least partially, supplant the relationship-based approach that I generally advocate in this book (Section 8.1).

1.8 Elements and Measures of Social Structure

Although within an ethological approach emergent properties are not the starting points of social analysis, they are highly significant. Elements of all three levels of Hinde's conceptual framework can often be measured or classified in useful ways and then used to investigate the evolution or function of social behavior or social structures as well as other questions. Interactions may be classifiable into agonistic, cooperative, or other classes, and the relative rates of the different classes might, for instance, be used as indices of the competitive or cooperative nature of a society (Section 5.1).

Moving up a level, the interactions between a pair of animals may have a particular content, quality, and patterning, allowing their dyadic

Table 1.2 Elements, Attributes, and Measures of Different Levels of Social Analysis

Attributes of individuals	Gender
	Age
	Reproductive state
	Role
	Gregariousness
	Dominance rank
Types of interaction	Agonistic
	Cooperative
	Reciprocal–unidirectional
	Symmetric–asymmetric
Types of relationship	Bonds
	Dependence
	Dominance
	Kinship
Elements within social structures	Groups
	Dominance hierarcies
	Roles
	Social units
	Tiers
Measures of social structure	Demographic structure
	Rates of interaction and communication
	Group size: number of potential interactants at any time
	Closure and stability of groups, units, or communities
	Stability of relationships
	Social differentiation: variability in probabilities of association (true association indices) among dyads
	Behavioral integration
	Patterns of communication
	Differentiation of roles
	Time and energy devoted to social behavior
	Community size: number of potential interactants over a substantial period of time
	Network measures such as mean and maximum path length, clustering coefficient, assortativity
	Social complexity

relationship to be classified (Section 4.8). For instance, some relationships may be dependent (e.g., parent–offspring), dominant (some types of interaction are consistently unidirectional), a bond (consistently strong in several independent modes), or refer to cooperative reproduction (such as a breeding pair). Types or degrees of relationship may also correlate with the attributes of the dyads, such as age or sex class or degree of kinship. Sometimes, relationships can be organized into mutually exclusive and/or hierarchically nested tiers, patterns of reciprocity, or dominance hierarchies. All these are elements of social structure (Table 1.2).

We can measure or classify social structures themselves. As noted earlier (Section 1.4) and by Wilson (1975, p. 16), classification has only been satisfactory within subsets of animal societies (Section 6.2). Measuring social structure is one of the aims of this book. To set the scene, here are Wilson's (1975, pp. 16–18) 10 "qualities of sociality," with some comments and cross-references to later sections of the book where they are considered:

· Group size. Group size is often the most obvious feature of an animal society. Surely, social life is different for an animal who can usually only sense one conspecific compared to an animal immersed in a group of hundreds. Defining and distinguishing animal groups is not straightforward, however, and groups may mean different things to different observers and to the animals themselves. For primatologists, a group is usually a largely self-contained set of animals whose interactions are principally with each other (what I call a community), whereas for scientists studying ungulates, a group may mean animals clustered for just a few minutes (e.g., Underwood 1981). In this book, I use *group* in the sense of animals that actively achieve or maintain spatiotemporal proximity, giving it a similar connotation to the primatologists' "party" but contrasting it with "aggregations" of animals that may be caused by nonsocial factors such as a patchy distribution of resources (Section 3.4). This definition of group is useful if associations are to be defined using groups, as is often the case (Section 3.4). Group size is an important measure of the complexity of social life, indicating the mean number of other individuals that an animal may interact with at any instant. Similarly, the community size (the primatologists' "group" size) is the number of other individuals with which an animal may interact over a substantial period of time, perhaps a season or a year.

· Demographic distributions. Age, sex, and reproductive state are fundamental characteristics of animals, strongly affecting their interactions and associations. With demographic information incorporated, descriptions of animal societies become much richer (Section 3.6).

· Cohesiveness. Wilson (1975, p. 16) suggests that the closeness of group members to one another may be an index of sociality. Fundamentally, we should be interested in rates of interaction, which are likely well predicted by cohesiveness and proximity.

· Amount and pattern of connectedness in communication. Patterns of communication within populations are an important feature of any social structure because communication is fundamental to sociality (Fig. 1.2). Network analysis addresses this (Section 5.3).

· Permeability. Wilson considers permeability in reference to movement between fairly closed groups. Permeability, in this sense, can be seen as one potential element of the temporal change in association, or interaction, patterns between dyads. These rates of change can be studied using measures such as lagged association rates (Whitehead 1995; Sections 4.6 and 5.5).

· Compartmentalization or modularity. The degree to which sets of animals operate as distinct units is an important element of social structure and can be examined using sociograms, cluster analyses, network analyses, and other methods (Section 5.7). More fundamentally, we can ask whether there are preferential associations among dyads (Bejder et al. 1998; Section 4.9).

· Differentiation of roles. The differentiation of roles is one of the significant attributes of social evolution in insects (Michener 1969). Among vertebrates, roles are less striking but may be present and important (e.g., Stander 1992). We can examine individual roles using multivariate analysis of interaction or association rates, network analysis, or other methods (Section 7.1).

· Integration of behavior. This is in some ways the obverse of the differentiation of roles and can be looked at using similar methods, although there are operational difficulties because lack of differentiation is often the null hypothesis, which cannot be proved. Measures of synchrony (Section 3.2) can be used, however, to examine integration from a more positive perspective.

· Rates of information flow. Wilson (1975, p. 18) suggests that we can usefully compare rates of information flow in societies of different types. There are many pitfalls here (some noted by Wilson 1975, pp. 199–200), however, including deciding what information is important to the sender and receiver, redundancy, how to compartmentalize information (e.g., into bits), and how to standardize time scales. I do not cover these areas in this book, but Bradbury and Vehrencamp (1998, esp. pp. 387–418) provide an excellent introduction.

- Fraction of time devoted to social behavior. This useful measure of sociality is one product of an ethogram that is described by Lehner (1998, pp. 90–93) but is not covered in this book. The proportion of energy devoted to social behavior would also be of interest.

Wilson's measures can be assessed and compared for populations of the same and different species and sometimes between classes of animal (such as those defined by age or sex) or even among individuals.

Missing from Wilson's list is what I think of as the grail of social analysis: a measure of social complexity that can be employed across species. We have not yet found such a measure. In Section 6.3, I discuss the desired characteristics of such a measure and prospects for success.

I have used most of the elements of Wilson's list and some other ideas discussed in this section to tabulate attributes, elements, and measures of the levels of social analysis in Table 1.2, which abstracts some of the principal elements and outputs of the methods discussed in this book.

1.9 The Functional Why Questions, and Ecology

Behavioral ecology is principally driven by functional questions (Krebs & Davies 1991). How do particular forms of behavior contribute to individuals' survival and reproductive success and thus become adaptive, and so persist? These questions are behind many, perhaps most, studies of animal societies; we are interested in the functional value of elements of a social structure to its members relative to other possibilities. In behavioral ecology, there are three principal methodological approaches: comparisons of individuals, experiments, and comparisons of species (Krebs & Davies 1991). Experiments are quite rare in vertebrate social analysis but are sometimes very illustrative (e.g., Kummer et al. 1974; Chase et al. 2002; Durrell et al. 2004; Flack et al. 2006). Instead, most frequently, we examine functionality in social systems by considering the position of individuals within a social structure and how patterns of interactions and relationships may relate to fitness. This is often done by relating nonsocial attributes of individuals and dyads, such as kinship and reproductive success, to measures of interaction or relationship. As discussed in Chapter 7, comparisons among communities, populations, and species may also be informative (e.g., Kappeler & van Schaik 2002).

Thus, from the behavioral ecologist's perspective, we may ask whether a particular interaction or relationship, or pattern of interactions or relationships, is adaptive to individuals, and if so, how. Constraints, such as a position in a dominance hierarchy, will often be important factors in such analyses. One might also consider the adaptive value of a particular social structure to its members, although in most cases this has no utility from an evolutionary perspective because social structure is an emergent property and individuals can do little to change the one they are in (Section 7.4).

A frequently employed approach is to look for ecological correlates of social structure. The assumption, usually unstated, is that factors such as population density, predation pressure, or the abundance and distribution of resources influence the patterns of interactions and associations among individuals and so their relationships and social structure. Then differences in these ecological factors can be related to differences in social structure. In this way, ecological factors explaining intraspecific, and especially interspecific, differences in social structure have been sought by many (including Crook & Gartlan 1966; Jarman 1974; Seghers 1974; Pitcher 1986; Wrangham & Rubenstein 1986).

1.10 Examples of Social Analyses

Although Hinde's framework is applicable to any population of animals, challenges and possibilities vary dramatically by species and population. In this book, I use examples from a range of species to illustrate methods. These can be arranged roughly along one major axis.

At one end are large populations of widely ranging animals in which particular individuals are encountered rarely and opportunistically, sexes and ages may not be known, and little behavior is observable. The work of my group on South Pacific sperm whales (*Physeter macrocephalus*), in which we opportunistically encounter members of a large population in a vast ocean area, is an example. The data collected are sparse and simple. Only a limited range of methods can be applied and a few topics examined—issues such as group size and the stability of associations. However, more can usually be retrieved from such data than might be imagined. It is important to remember that because we cannot see or measure much social behavior in such cases, this does not mean that social structure is simple.

At the other extreme are small, easily viewed captive colonies (e.g., Corradino 1990). Animals are well known to the scientists and usually each other, not only their ages, sexes, and kinship, but also their social histories. The range of options for social analysis with such populations is huge, including the analysis of shifting dominance hierarchies and

Table 1.3 Some Studies of Vertebrate Social Structure, Arranged in Order of the Number of Individuals Identified

Species	Days observed	Span of study area	Identification used	Identified animals	Questions asked	Reference
Black-capped chickadee, *Parus atricapillus*	~30	~20 m[a]	Color banded	9	Social affiliations, aggression, dominance, and pairing within a flock	Ficken et al. 1981
Japanese monkey, *Macaca fuscata fuscata*	11.4	~50 m[a,b]	Natural markings	14	Individual and class variance in sociality and proximity	Corradino 1990
Domestic pig	42	2 × 4 m pens[a,b]	Spray paint marking	33	Do pigs form long-term preferential associations? (experimental manipulation to eliminate spatial preferences)	Durrell et al. 2004
Sanderling, *Calidris alba*	40	4 km	Colored leg bands	~48	Do individuals form flocks randomly?	Myers 1983
Mouflon, *Ovis orientalis*	~200	7 km[a]	Collars and ear tags	62	Group size, ranging, social preferences by age/sex class	Le Pendu et al. 1995
Northern bottlenose whale, *Hyperoodon ampullatus*	96	50 km	Photos of natural markings	65	Duration of association patterns among individuals within and between age classes	Gowans et al. 2001
Cichlid fish, *Julidochromis marlieri*	11	20 m	Natural markings	67	Describe social structure, ranging, nesting	Sunobe 2000
Chimpanzee, *Pan troglodytes*	1,000s	15 km[a]	Natural markings	137	Describe interactions, relationships, aggression, grooming, dominance, others	Goodall 1986
Disk-winged bat, *Thyroptera tricolor*	42	500 m	Forearm bands	336	How do roosting companionships interact with ranging patterns?	Vonhof et al. 2004
Sperm whale, *Physeter macrocephalus*	244	200 km	Photos of natural markings	1,767	Describe hierarchically nested tiers of social structure among females and interactions with breeding males	Whitehead 2003

[a] Partially or fully provisioned.
[b] Captive populations.

triadic interactions. The disadvantages are that the habitat is unnatural and that the social structures of small populations can be strongly influenced by particular individuals and events (for a dramatic example see de Waal [1998]), so generalization becomes problematic.

Table 1.3 lists 10 diverse studies of vertebrate social structure published over the last 25 years. It includes birds, fish, and mammals ranging in size from chickadees to sperm whales. There are aquatic, aerial, and terrestrial species. These studies took place on the lands or waters of nine nations and examined free-living, provisioned, and captive animals. In one case, there was an experimental component to the study. The studies ranged very considerably in scale, with 9 to 1,767 animals being identified individually through a variety of methods over spatial spans ranging from 4 m to 200 km. They also ranged very considerably in scope, from work directed toward one principal question, such as whether individuals associate randomly (Myers 1983), to multifaceted and detailed descriptions of social life (Goodall 1986). It is clear that to look more deeply, one needs more data. The numbers in Table 1.3 suggest that in most circumstances about 30 study days are needed to make an informative study of social structure. However, snorkeling or using scuba equipment allows some fish societies to be studied very efficiently, with useful descriptions of social structure emerging from just a few days of study. The power and precision of social analyses are considered in Section 3.11.

Despite their variety, there are some communalities among the studies in Table 1.3 that reflect the thrust of this book. In all studies, animals were individually identified (Section 3.5), although sometimes unidentified members of the same population were also present, and in all studies there was an attempt to describe social structure as the pattern of relationships among members of the population (Chapters 4 and 5).

1.11 Problems with Analyzing Social Structure

Social analysis is rarely straightforward. If it were, there would be no need for this book. Scientists are faced with a number of questions, challenges, and decisions. Most fundamental is the research goal. In Chapters 3 to 5, I usually assume a very general objective: describing the social structure of an animal population with maximum precision. The methods considered, however, should be useful when only one element of social structure is being examined or a hypothesis, causal or functional, is being tested (Chapter 7). In these cases, it is very important to state the question clearly and design the data collection and analysis appropriately.

Thus, before starting research, we should consider some fundamental questions. Which data should be collected? How much data are needed to build a "useful" model of social structure or investigate our questions? In reality, the data have often been collected before the analysis is properly considered, and sometimes primarily for some quite different purpose, such as population analysis. In such circumstances, we should do the best with what we have but not try to build models or come to conclusions that go beyond the power of the data.

With data in hand, or, better, in a database or spreadsheet, another set of questions needs attention before the numbers are crunched. Which data can be used? Are the data independent, and, if not, is this a problem? How do we study social structure in the many cases in which interactions, the basis of Hinde's framework, are not visible? Which analytical methods, which options, and which computer programs should we use? Should hypotheses be tested or models fitted? If hypotheses are considered, which null models are appropriate? If we are fitting models, which should be tried?

As the results appear, we still need to question. Does the output make sense when compared with what we know of the animals and the data set? How can we distinguish real features of animal social structures from methodological artifacts of the data?

And then there are more general issues. How can rare but vital events such as weaning and mating be incorporated into social analysis? How may social structures of different populations or species be compared?

In this book I try to provide guidance on these issues.

2

Technical Matters

The principal function of this chapter is to introduce some standard statistical methods and terms (summarized in Table 2.1) that can be useful in several areas of social analysis. More-specific methods are presented later in relevant sections. I begin the chapter with a more general and fundamental discussion of the ways in which we do science and end it with a summary of computer packages that can assist with social analysis.

Social data are different—different from the majority of the data in introductory statistics classes or textbooks (e.g., Sokal & Rohlf 1994; Ruxton & Colegrave 2006). This is because they usually refer to the interactions, associations, or relationships between two (or occasionally more) individuals (Hinde 1976)—they are dyadic. This means that they can rarely, as they stand, be considered independent, and should not be subject to statistical methods, such as t-tests, that assume independence. Instead, we need more-specialized techniques, such as ordinations, permutation tests, and the class of methods termed network analysis.

2.1 Modes of Scientific Enquiry

The foundation of the scientific method in biology is usually considered to be the designed manipulative experiment

Table 2.1 Some Standard Statistical Methods Useful in Social Analysis

Basic statistics (section 2.2)	Measures of central tendency	Mean, median, mode
	Measures of dispersion	Variance, standard deviation, coefficient of variation, range
	Relationship between variables	Covariance, correlation coefficient, Spearman rank correlation coefficient
Precision (section 2.3)	Precision of statistic or parameter estimate	Standard error, confidence interval from: distributional probability functions, likelihood methods, parametric bootstrap, nonparametric bootstrap, jackknife
Hypothesis testing (section 2.4)	Traditional techniques	t-test, ANOVA, MANOVA, others
	Permutation tests	Mantel tests, others
Data matrices (section 2.5)	Rectangular data matrices	Individual by attribute, group by individual, others
	Similarity and dissimilarity matrices	Matrices of association indices, matrices of kinship, others
Ordination (section 2.6)	Visual representation of data matrices	Principal components analysis, correspondence analysis, principal coordinates analysis, nonmetric multidimensional scaling
Classification (section 2.7)	Nonhierarchical cluster analysis	K-means, network methods of community delineation
	Hierarchical cluster analysis	Average linkage, Ward's, others
Model fitting (section 2.8)	Parameter estimation, model selection	Maximum likelihood, Akaike information criterion

(Ruxton & Colegrave 2006). The scientist comes up with a hypothesis, usually either from theory, a preliminary study, or results from a related system. The hypothesis is often phrased in terms of some factor or factors having an effect on a system. Then, in the simplest form, the factor is randomly applied or not applied to a set of independent subjects and an output variable is measured. The scientist uses statistical methods to accept or reject the null hypothesis that the factor has an effect (e.g., Lehner 1998, pp. 347–357; Martin & Bateson 2007, p. 106). This methodology can be elaborated in many ways, but the essence remains testing whether factors affect a biological system by controlling their application.

In the analysis of the social systems of free-ranging vertebrates, designed experiments are rare. For many species, experiments that would reveal important aspects of social structure are either impossible or sufficiently difficult, costly, or ethically questionable that they are not carried out. Instead, much use is made of quasi-experiments in which natural variation in independent variables is used to investigate their relationship with social measures. For instance, a scientist might compare group size with predator density across study sites to investigate the hypothesis that the function of groups is in defense against predators (e.g., Seghers 1974). Quasi-experiments are not as definitive as real manipulative experiments. A statistically significant positive correlation between group size and predator density among study sites seems to support the defensive function hypothesis, but it could also be due to the effects of some confounding variable. Perhaps more-productive sites can support both larger groups, whose primary function is in intraspecific competition, and more predators.

Often social analysis is purely descriptive: We wish to produce a model of social structure that matches reality as closely as possible, given the effort we can put into collecting data or the data available. In producing such models, it is often useful to have formal or informal hypotheses about the social system that we can use to guide the data collection and analysis (e.g., that there are *social units* with permanent, or nearly permanent, membership, or that bonds among members of one sex are stronger than those among members of the other), and these may be developed into statistically analyzed quasi-experiments to examine the relationships between the independent variables and the social measures, assuming no unaccounted-for confounding variables. Unlike many of those who write about biological data analysis, however, I do not believe that hypotheses are always necessary. Effective social analyses, especially of poorly known species, can often proceed without hypotheses: General principles of data collection are used to collect a potentially revealing data set, which is then displayed using hypothesis-free analytical methods such as ordinations, sociograms, or lagged association rates. The patterns that emerge may suggest hypotheses that can be tested using aspects of the original data that are independent of the exploratory analysis or, better, a new data set. The testing of hypotheses on the data that were used to develop them is almost always wrong (e.g., Burnham & Anderson 2002, pp. 37–38).

As noted, hypothesis testing is not as common in social analysis as in some other areas of biology. However, many have argued (e.g., Johnson 1999) that biologists test hypotheses for statistical significance *too* often, and that many null hypotheses are straw men, most unlikely to be true. When investigating sex differences, for instance, the size of the effect—the magnitude of the effect of a factor on the response variable, in this case sexual dimorphism—is usually the important issue, not whether it is statistically significant. Thus the presentation of effect sizes, and confidence in them, should take precedence over the P values of hypothesis tests (Johnson 1999).

In social analysis, however, there are reasonable null hypotheses, and falsifying them should precede some further analyses. As an example, it is entirely conceivable that individuals have no preference as to the identity of their associates (e.g., if there is no means of individual recognition), and if this is so, displays of patterns of association among individuals (such as dendrograms) have little validity.

Unfortunately, testing hypotheses of social patterning is not particularly straightforward. Social measures, such as association indices, are rarely normally distributed, and more important, they have patterns of dependence that violate the assumptions of most standard statistical tests. For instance, the association index between individuals A and C is usually not independent of the indices between A and B and between B and C. To accommodate these issues, we generally use permutation tests, but we need to set them up carefully. Later (e.g., Section 4.9), I will show how to set up these types of hypothesis tests.

I will also use another mode of statistical enquiry—model fitting. Here, the assumption is that the data collected are produced by a complex process and that we wish to find a mathematical model that best approximates it. The best model balances bias versus precision (Burnham & Anderson 2002, p. 32). A simple model may have bias in that it consistently errs in its predictions at some combinations of parameter values; a complex model will have imprecise parameter estimates.

2.2 Basic Descriptive Statistics

Let us suppose that we observe n subjects (often individual animals in social analysis) and, on each, measure variable x, so we have an n-element data set $\{x_1, x_2, \ldots, x_n\}$. In social analysis, x might represent the rate of some kind of display per time unit. We can use a range of "statistics" to describe these data (see Sokal and Rohlf [1994, pp. 39–59]) for more information]. Basically, there are three well-used measures of central tendency:

> *Mean.* The sum of observations divided by the number of observations ($\bar{x} = \Sigma x_i / n$). The mean is the most frequently used measure of central tendency.
>
> *Median.* The value above and below which there are equal numbers of observations. The median is particularly useful in situations in which measures have unusual distributions or there are outliers.

Mode. The most common value of a variable, the peak value of
the frequency distribution. A bimodal distribution has two
peaks, a multimodal distribution many peaks, and so on.

There are several measures of dispersion of a variable about its
central tendency:

Variance. The mean square deviation of the observations from
the mean: $V = \Sigma(x_i - \bar{x})^2/n$. This is the variance of the
sample. The variance of the original variable is better estimated by the following formula, which removes bias:
$V = \Sigma(x_i - \bar{x})^2/(n-1)$. The variance is theoretically very
important but hard to relate to actual phenomena because
units are squared. Instead we use the following:

Standard deviation (SD). $SD = \sqrt{V}$. The standard deviation is
the most common measure of dispersion. It is in the same
units as the original variable. With a normal distribution,
about 70% of the observations are within one SD of the
mean, and about 95% are within two SDs of the mean.

Coefficient of variation (CV). $CV = SD/\bar{x}$. The coefficient of
variation is a standardized, unit-free measure of dispersion.

Range. The difference between the minimum and maximum
observations. The range is influenced by sample size.

If, for each individual, we measure two variables x and y, then we
can describe the relationship between them in several ways, including
the following:

Covariance. The equivalent of the variance for two variables,
usually estimated from $C(x, y) = \Sigma(x_i - \bar{x}) \cdot (y_i - \bar{y})(n-1)$.
The covariance is large and positive if x and y are positively
related to one another, large and negative if they are inversely related to one another, and close to zero if they are
unrelated. However, the units of covariance (the product of
the units of the two original variables) are not particularly
useful. Thus, instead, in descriptive statistics, we use the
following:

Correlation coefficient. The correlation coefficient is the covariance standardized by the standard deviations of the
two measures: $r = C(x, y)/[SD(x) \cdot SD(y)]$. The correlation

coefficient has no units and ranges between $+1$, indicating a positive linear relationship between the variables, and -1, indicating a negative linear relationship between the variables. A value of $r = 0$ indicates no relationship.

Spearman correlation coefficient r_s. Before computing the correlation coefficient, the two variables are ranked, so the lowest x becomes 1, the next lowest 2, and so on, and the lowest y becomes 1, the next lowest 2, and so on. The Spearman correlation coefficient is useful in situations in which the relationship between x and y may not be linear. The value of r_s also varies between -1 and $+1$, with $r_s = 0$ indicating no relationship.

2.3 Precision of Statistics: Bootstraps and Jackknives

An estimated social measure (such as mean group size) or parameter (such as rate of disassociation) may be a useful output of a social analysis, but it has little value without some measure of precision. Estimates of precision are often presented using the statistics of dispersion summarized in the previous section. Instead of describing how a set of data is dispersed around its mean or median, however, they are used to quantify the error between an estimated parameter or statistic and its true value. Standard deviations of statistics are often known as standard errors (SE), and these are commonly used to describe confidence in an estimate. Coefficients of variation (in this case $CV = SE/\bar{x}$) can also be appropriate.

The other common way of expressing the precision of an estimate is through a confidence interval. For a single-valued statistic, two numbers make up the interval such that if we repeatedly sampled the population in the same manner and calculated confidence intervals, some percentage—usually 95%—of these intervals would contain the true mean.

In some circumstances, estimating standard errors and confidence intervals is straightforward. For instance, the usual estimated standard error of the mean of a population calculated using a sample of size n, $\{x_1, x_2, \ldots, x_n\}$, is simply $SD(x_i)/\sqrt{n}$, and its 95% confidence interval is

$$[\bar{x} - t_{0.05[n-1]} \cdot SD(x_i)/\sqrt{n}] \text{ to } [\bar{x} + t_{0.05[n-1]} \cdot SD(x_i)/\sqrt{n}]$$

where $t_{0.05[n-1]}$ is the two-sided cumulative probability distribution of the t distribution with $n - 1$ degrees of freedom, which is often tabulated at the back of introductory books on statistics. As a rough guide, the 95% confidence interval of a normally distributed statistic is the mean plus and minus about twice the SE.

Estimating the precision of other statistics and parameter estimates is less straightforward. Social measures are often like this. For instance, there is no easy formula for giving the precision of an estimate of the mean association index (Section 4.5) among a population of animals. Techniques exist, however, that can be used in difficult situations like this. Likelihood methods (Edwards 1992; see later discussion) can be used to estimate model parameters, and likelihood also gives estimates of precision, although the mathematics is sometimes difficult. Two computer-intensive techniques have become very important in this area over the last two decades: the bootstrap and the jackknife (Efron & Gong 1983; Sokal & Rohlf 1994, pp. 820–825).

There are actually two, quite different forms of the bootstrap (Fig. 2.1). To use the parametric bootstrap to estimate parameter precision, we need a model of the process by which the parameters produce the data. Having estimated parameters from the real data, we feed them into the model to produce random data by Monte Carlo methods (Manly 1997, p. 61), from which we estimate a new set of parameters. This is a bootstrap replicate, and the distribution of many such bootstrap replicate parameter estimates tells us about the bias and precision of the original estimate from the real data. The parametric bootstrap is an important technique, giving, in some respects, the best-possible estimate of precision, as well as other important information about the estimation procedure. However, it takes some skill in computer programming to achieve (because the generative model must be coded), and is infrequently used in practice.

In contrast, the nonparametric bootstrap—usually, and hereafter in this book, just referred to as "the bootstrap"—is quite easy to use and has become a preferred method of estimating precision in many areas of research (Fig. 2.1). In the nonparametric bootstrap, we assume that the original data consist of n independent units—they must be independent. Having estimated the parameter(s) or statistic(s) using the data set, we construct bootstrap replicates sampling with replacement from the n independent units in the original data set. Thus, in a single bootstrap replicate, x_1 might occur three times and x_2 not at all. From each bootstrap replicate, we calculate parameter estimates, as with the original data. For any parameter, the distribution of its estimates among, say, 1,000 bootstrap replicates allows precision to be estimated. The SD of the bootstrap replicate parameter estimates is used as the SE of the original

$$x_1, x_2, ..., x_n \Rightarrow \text{Statistic / Parameter estimate } \hat{s}$$

Parametric Bootstrap	*Non-Parametric Bootstrap*	*Jackknife*
{Assume model M: $M(s) \Rightarrow x$}	Sample from $\{x_1, x_2, ..., x_n\}$ with replacement to produce: $x_1^{(1)}, x_2^{(1)}, ..., x_n^{(1)}$	Omit x_1 to give: $x_2, ..., x_n$
Use M and $\hat{s}$ to produce random data: $x_1^{(1)}, x_2^{(1)}, ..., x_n^{(1)}$		Estimate statistic for data missing x_1: $\hat{s}(-1)$

Estimate statistic omitting each data point in turn : $\hat{s}(-1), \hat{s}(-2), ..., \hat{s}(-n)$

Estimate statistic for random data: $\hat{s}^{(1)}$

Estimate statistic for many sets of random data: $\hat{s}^{(1)}, \hat{s}^{(2)}, ..., \hat{s}^{(1000)}$

Convert to pseudovalues: $\phi_i = n \cdot \hat{s} - (n-1)\hat{s}(-i)$

Estimated SE($\hat{s}$) = SD($\hat{s}^{(1)}, \hat{s}^{(2)}, ..., \hat{s}^{(1000)}$)

Estimated SE($\hat{s}$) = SE($\phi_1, \phi_2, ..., \phi_n$)

Other measures of confidence in $\hat{s}$ come from distribution of: $\{\hat{s}^{(1)}, \hat{s}^{(2)}, ..., \hat{s}^{(1000)}\}$

FIGURE 2.1 Bootstraps and jackknives: the basics.

estimate, and its 95% confidence interval is such that 2.5% of the bootstrap replicate parameter estimates are less than the lower bound and 2.5% are above the upper bound. The bootstrap is a powerful and widely used technique, but it is invalid in situations in which having identical units in the data set (as results from sampling with replacement) biases the statistics or parameter estimates. This occurs in social analysis when the statistic or parameter reflects association among individuals. For instance, association indices between an individual and itself are always 1.0, so the bootstrap cannot be used to estimate the precision of an estimate of the mean association index among individuals within a population by randomly sampling the individuals with replacement for each bootstrap estimate.

In such cases, the jackknife (Fig. 2.1) is a useful alternative. Jackknife replicate data sets are calculated by omitting each unit (or set of units) in turn from the data set, and jackknife parameter estimates are obtained from these [$\hat{s}(-i)$ is the estimate omitting unit i]. Pseudovalues [$\varphi_i = n \cdot \hat{s} - (n-1) \cdot \hat{s}(-i)$, where n is the sample size and $\hat{s}$ is the original parameter estimate] are then calculated. The distribution of the pseudovalues gives estimates of the bias and precision of the original parameter estimate (Fig. 2.1). Jackknife estimates of precision tend to be conservative, overestimating standard errors and not very precise themselves (Efron & Stein 1981). However, the jackknife is useful in cases when other methods are not available. The data units that are sequentially omitted should be independent of one another. I have found jackknife techniques in which temporal clusters of data are omitted in turn (e.g., each month of collected data) to be useful in some situations when there is no other straightforward method of assessing precision.

2.4 Hypothesis Testing

As noted earlier (Section 2.1), formal hypothesis testing does not have as large a role in social analysis as in most other areas of biology. However, it has a role. We can sometimes set up realistic null and alternative hypotheses and use data to test them against one another. For instance, it might be reasonable to test the null hypothesis that groups including only males are of similar size to those containing only females against the alternative hypothesis that one sex forms larger groups than the other, or the hypothesis that rates of aggression are similar in groups of all sizes against the alternative that aggression rates rise with group size. In the latter example, we may be able to use classical experimental design to test the hypothesis: We can randomly select individuals from a population, without replacement, to construct experimental groups of various sizes; measure aggression rates for a randomly chosen individual in each group in a standard manner; and then test whether the correlation coefficient between group size and aggression rate is significantly greater than zero, using the well-known distribution of the calculated correlation coefficient under the null hypothesis of no relationship between the variables. Standard methods of hypothesis testing using designed experiments are described in many books, including those by Sokal and Rohlf (1994) and Ruxton and Colegrave (2006). Assumptions and issues with these standard statistical methods are discussed briefly in Chapter 7 in Box 7.1.

Frequently in social analysis, however, including the example of differences in group size with sex, manipulative experiments are impossible: We cannot randomly assign sex. Thus, as noted earlier, we are left with quasi-experiments: We take the sexes the animals come with and see whether males are in different-size groups than are females. With a quasi-experiment, confounding explanations must be carefully considered. Perhaps larger animals are usually found in smaller groups and males are larger, and so the bigger groups of females are a secondary result of sexual dimorphism. In this case, however, and frequently in social analysis, there is another problem preventing the employment of standard statistical techniques for testing hypotheses, in this case the t-test. We would normally make observations of a number of groups and record the genders present. Usually, there will be individuals in more than one of the observed groups, so the groups are not independent. For instance, a particularly asocial male, always alone, will bias the data. t-tests, ANOVAs, and so on, all assume independence of experimental subjects, and when independence is lacking, test results are invalid. What to do? The best answer in many cases is to use permutation tests.

2.4.1: Permutation Tests. In permutation tests (Manly 1997), a test statistic is chosen and calculated for the real data. The data are randomly permuted—scrambled—many times (using a computer), with the statistic being calculated for each of these permuted data sets. The real statistic is compared with those from the randomly scrambled data sets, and if it is, say, less than or greater than those from 97.5% of the randomly permuted data sets, the two-sided null hypothesis that the permutation had no effect is rejected (at $P = 0.05$). Sometimes we must be careful in translating this result into biologically meaningful terms—the hypothesis usually only refers to the data collected, and this method cannot be used as it stands to test absolute values of parameters of populations against a null value (Manly 1997, p. 17). Thus, in the case of differences in group size with sex, we might proceed by taking all records of group membership and calculating the difference between the mean size of the groups only containing males and those only containing females. Then, for each permutation, sexes are scrambled so that there are the same number of males and females as in the real data but individuals may be of the opposite sex. Groups of mixed sex are then discarded and the difference in mean group sizes between those containing only males and those containing only females is calculated, giving one random value of the test statistic. Suppose this statistic is greater in the real data than in 993 of 1,000 permuted data sets; we can conclude that, within the data set, males form larger groups than females with a two-sided significance $P = 0.014$.

Permutation tests are discussed in some detail by Manly (1997). Technical issues that need considering include how many permutations to carry out. In the usual situations in which permutations are independent, 1,000 permutations are usually sufficient, although more may be used if the P value is close to a critical value. There are some permutation tests, including one that is very important in social analysis (Bejder et al. 1998; Section 4.9), in which the permutations are not independent of one another. In such cases, many more permutations are usually required. Several computer packages can be used to carry out permutation tests (Appendix 9.3).

A relatively simple and important permutation test is the *Mantel test* (Mantel 1967; Schnell et al. 1985). This examines the relationship between two similarity (or dissimilarity) matrices (Section 2.5). The matrices are indexed by the same subjects, usually individuals in social analysis. Thus, we might have a similarity matrix giving the rates of association among members of a community and another with their genetic relatedness (e.g., Table 7.1). A *matrix correlation* can be computed between the two matrices: it is simply the correlation coefficient between the elements of the different matrices, ignoring those on the diagonal. This indicates the degree of association between the two measures, with $r > 0$ indicating that higher genetic relatedness predicts higher association indices. Because of patterns of dependency (each individual contributes to many association and kinship values), however, the null hypothesis of no correlation between association and kinship cannot be tested using the standard formulas and tables for confidence intervals of r under the null hypothesis that its true value is zero. Instead, the Mantel test permutes the identities of the individuals in one matrix many times, calculating the matrix correlation coefficient for each permutation. The true value of r is compared with the distribution of these random r's to investigate its statistical significance. There are variants on the Mantel test, including nonparametric versions that consider the ranks of elements of the similarity matrix (Hemelrijk 1990a) and partial Mantel tests in which the relationship between two similarities is controlled for a third (e.g., whether grooming rates are related to association indices after controlling for difference in dominance rank) (Smouse et al. 1986; Hemelrijk 1990a).

2.5 Data Matrices

Social data are usually represented by matrices (two-way tables of numbers). There are two principal forms.

The simplest form is the rectangular data matrix. It is indexed by subjects (by convention giving the rows) and variables (the columns). An important example is a group-by-individual matrix. Rows represent groups and columns represent individuals, with "1" indicating that the individual was in the group and "0" that it was not in the group, or perhaps that it was not observed in the group. So we have something like the matrix in Table 2.2.

Table 2.2 Example of Group-by-Individual Matrix

	Individual				
Group	JOE	SAL	FRED	BOB	SUE
a	1	0	0	0	1
b	0	1	1	0	1
c	0	1	1	0	1
d	0	1	0	1	1
e	1	1	0	1	0
f	0	0	0	1	0
g	0	0	1	0	0
h	0	1	1	0	1

Each entry indicates whether an individual was present (1) or absent (0) in a particular observed group.

Table 2.3 Example of Similarity Matrix between Individuals.

	JOE	SAL	FRED	BOB	SUE	CON	ART	BILL
JOE	1	0	0	0	1	0	0	0
SAL	0	1	1	0	1	0	0	1
FRED	0	1	1	0	1	0	1	1
BOB	0	0	0	1	1	1	0	0
SUE	1	1	1	1	1	0	0	1
CON	0	0	0	1	0	1	0	1
ART	0	0	1	0	0	0	1	1
BILL	0	1	1	0	1	1	1	1

Each entry indicates whether a dyad was associated (1) or not (0) during a sampling period. This matrix is symmetric.

The second form of data matrix frequently used in social analysis is a similarity matrix. Here, both the rows and columns generally refer to individuals, usually the same set of individuals. Sociologists call this a "sociomatrix." Similarity matrices are useful in several situations. A symmetric 1:0 matrix can represent associations within a sampling period: a "1" in the cell at the intersection of A's row and B's column indicates that A and B were associated; a "0" in the same cell indicates that they were not associated using some definition of association (e.g., proximity, behavioral synchrony; Section 3.3). It is symmetric in the sense that this value is the same as that in the cell at the intersection of A's column and B's row, so that if A is associated with B, then B is associated with A, as in Table 2.3.

A similarity matrix need not be symmetric, for instance, if association is defined in terms of nearest neighbor distances, so that A may be B's nearest neighbor but B is not A's. Similarity matrices can show integer (i.e., 0, 1, 2,...) counts of dyadic interaction events. These are sometimes called sociometric matrices (Lehner 1998, p. 201). Such

Table 2.4 An Asymmetric Similarity Matrix Giving Frequencies of Occurrence, Perhaps Representing Numbers of Contests Won during Dyadic Agonistic Encounters

	JOE	SAL	FRED	BOB	SUE	CON	ART	BILL
JOE	0	8	4	2	6	1	0	3
SAL	0	0	1	7	3	2	1	1
FRED	9	1	0	0	5	0	2	1
BOB	0	0	0	0	1	1	0	0
SUE	1	4	1	1	0	0	6	1
CON	0	3	3	0	0	0	9	1
ART	0	0	1	0	0	0	0	1
BILL	7	5	9	6	2	2	1	0

Table 2.5 Symmetric Similarity Matrix of Association Indices Using Half-Weight Association Indices (section 4.5) of Observations of Members of a Social Unit of Seven Sperm Whales (*Physeter macrocephalus*) Observed off Dominica in 2005

	#5703(C)	#5722(A)	#5561(A)	#5727(J)	#5560(A)	#5130(A)	#5563(A)
#5703(C)	1	0.97	0.48	0.35	0.29	0.13	0.13
#5722(A)	0.97	1	0.17	0.06	0.12	0.04	0.07
#5561(A)	0.48	0.17	1	0.32	0.29	0.27	0.33
#5727(J)	0.35	0.06	0.32	1	0.53	0.11	0.07
#5560(A)	0.29	0.12	0.29	0.53	1	0.18	0.07
#5130(A)	0.13	0.04	0.27	0.11	0.18	1	0.09
#5563(A)	0.13	0.07	0.33	0.07	0.07	0.09	1

Data from Gero (2005). A, adult female; C, male calf; J, juvenile male. The numbers indicate the proportion of time that each pair of individuals was clustered together when at the surface.

matrices may be symmetric (e.g., if the event is touching) or asymmetric (e.g., if it details grooming or agonistic behavior), as in the example of Table 2.4.

Similarity matrices representing associations or behavior can have noninteger values, for instance, giving mean interindividual distances. Frequently, several similarity matrices are needed to represent social data, one for each type of behavior or association measured and one set of these for each temporal sampling period.

Similarity matrices can also be used to represent the results of social analyses. In particular, association indices between pairs of individuals within a population can be tabulated in similarity matrices (e.g., Table 2.5). Association indices estimate the proportion of time that pairs of individuals are associated (Section 4.5).

A dissimilarity matrix is structured like a similarity matrix, with individuals as both rows and columns, but here, high values indicate less similarity. A matrix of geographic distances between locations is a dissimilarity matrix. Dissimilarity matrices can be converted into similarity matrices by simple transformations, such as "$(1 - x)$" or "$(1/x)$". Association indices, and so matrices of association indices, are usually transformed into dissimilarities by taking the square root of 1 minus the association index.

Sometimes similarity and dissimilarity matrices are lumped under the term "association matrix." Using "association matrix" in this way would be confusing in social analysis because "association" has an important meaning of its own (Section 3.3), and matrices of association indices are commonly used.

2.6 Ordination

Human brains are not particularly good at assimilating tables of numbers, such as those in a group-by-individual matrix or a similarity matrix. A number of techniques have been developed to ordinate such data to display them visually in ways that we can assimilate much better. *Ordination* tries to represent subjects as points in a low-dimensional visual display (Randerson 1993). There are many examples of ordination in this book, usually with individuals being represented by points in a two-dimensional display, with the proximity of the points indicating their association or relationship. Several ordination methods are available [for more information see, e.g., Randerson (1993), Manly (1994), and Legendre and Legendre (1998)]. Four of the most important techniques are as follows:

> *Principal components analysis.* This, the most basic ordination method, works with a rectangular data matrix, preferably with normally distributed elements. Such data matrices do not occur very frequently in social analyses (but see, e.g., Section 5.6 and Fig. 5.18).
>
> *Correspondence analysis.* This technique is better adapted than principal components analysis for ordinating data matrices with 1:0 or positive integer elements, such as group-by-individual matrices. An ordination using correspondence analysis of the group-by-individual data matrix in Table 2.2 is shown in Fig. 2.2. Note that SUE and SAL, which were together in four of five groups in which each was observed, are plotted together, whereas BOB, with only one group in common with any of the other individuals, is plotted separately. Another example is shown in Fig. 7.1, which illustrates the

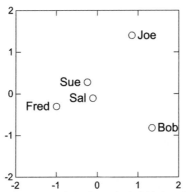

FIGURE 2.2 Example of correspondence analysis of group-by-individual matrix (table 2.2). Individuals generally identified in the same groups are plotted together.

ability of correspondence analysis to visually display both the rows and columns of a data matrix in one figure.

Principal coordinates analysis (also called *classic metric multidimensional scaling*). This is a modification of principal components analysis that can be used on similarity matrices, such as matrices of association indices (Digby & Kempton 1987, pp. 83–93). Although in most applications, principal coordinates analysis actually uses dissimilarity matrices (in which large values indicate unlike subjects), it can employ similarity matrices by a suitable transformation of association into dissimilarity (such as the square root of 1 minus the similarity, the common transformation in social analyses). It tries to produce a plot in which the distance between two points on the plot is proportional to the dissimilarity between the two individuals. A principal coordinates ordination of the matrix of association indices of sperm whales in Table 2.5 is shown in the upper part of Fig. 2.3. Three clusters appear: the calf and one adult female (its mother), the juvenile and another adult female (its mother), and then the other three adult females (Gero 2005). More information and additional examples of principal coordinates analysis are given in Section 5.2.

Nonmetric multidimensional scaling (MDSCAL). This method also uses a similarity matrix and produces a plot similar to that of principal coordinates, although, unlike principal coordinates, it can use a similarity matrix directly without

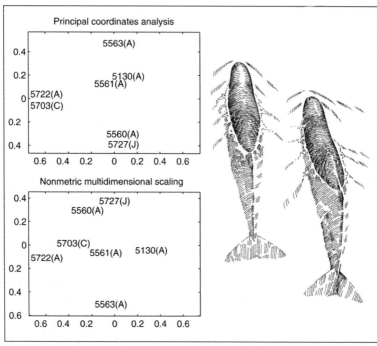

FIGURE 2.3 Example of principal coordinates analysis and nonmetric multidimensional scaling of the matrix of association indices of the social unit of sperm whales in table 2.5. (Illustration copyright Emese Kazár.)

transformation into dissimilarities. The plots of the sperm whale data (Table 2.5) in Fig. 2.3 using the two methods have some similarities (the orientation of these plots is arbitrary), although the clustering in the multidimensional scaling plot is much less tight, with the calf taking a more central position. MDSCAL differs from principal coordinates analysis in several ways. The relationship between the associations and the distances between the points tries to be monotonic (i.e., more associated pairs are represented by closer points) rather than linear. This is a less stringent criterion, so with MDSCAL it is easier to ordinate data in few dimensions than with principal coordinates. MDSCAL is also an iterative procedure that may not always reach the same solution, in contrast to principal coordinates, which is prescriptive and gives a unique display. Unlike principal coordinates, there is a limit on the number of subjects that MDSCAL can ordinate (perhaps a few hundred, depending

on the computer package). In social analysis, a linear relationship between transformed similarities and distances on
the ordination will usually be less important than captur-
ing as much as possible of the general pattern of similarity,
so MDSCAL is usually preferable to principal coordinates
(Section 5.2).

I have stressed two-dimensional ordinations of data or similarity
matrices in this section. The principal output are "scores," the coordinates of the points in the two dimensions (as depicted, for instance
in Fig. 2.3). However, the ordination techniques can give much more
(Section 5.2). Ordination can also be done in one, three, or more dimensions. The methods give measures (including the "stress," proportion of
variance accounted for, etc.) of how well the data are represented in
any specified number of dimensions. When a data matrix is ordinated
using principal components or correspondence analysis, then we also
get useful information on the other axis of the matrix. In the case of a
group-by-individual matrix, these are groups, so the principal components analysis shows how they are associated with the ordination axes,
in other words, which individuals were associated with which groups.

2.7 Classification

Another set of methods for simplifying and displaying data matrices is
classification or cluster analysis, in which the subjects are assigned to clusters. In social analysis, the subjects are usually individuals and the
clusters may be permanent or semipermanent social units, communities,
or other social entities. Unlike the model-free ordinations, classification
is only appropriate when the society does contain such clusters. We can
classify random data, producing informative-looking clusters (e.g., Fig.
5.7), but they mean nothing, as do quite a few of the published classifications of individuals in studies of animal societies.

There is a large range of statistical methods available for classification. For introductions to cluster analysis see, for instance, Bridge
(1993), Legendre and Legendre (1998, pp. 303–385), and Manly (1994,
pp. 128–141). Cluster analyses can be divided into nonhierarchical and
hierarchical methods.

In nonhierarchical cluster analysis, a set of subjects (usually individuals in social analysis) is separated into clusters such that, in some defined
sense, the separation between the clusters is maximized. Two important, and challenging, questions are the following: How many clusters

should we use, and how should we measure intercluster separation? The standard method of nonhierarchical cluster analysis is K-means, which works using a data matrix, such as the individuals-by-groups matrix of Table 2.2. In K-means, subjects are allocated to a given number of clusters to minimize the pooled within-cluster sum of squares. K-means leaves open the question of how many clusters to use. For instance, using K-means on the data in Table 2.2 gives the following allocations:

Two clusters: {SAL, FRED, SUE}, {JOE, BOB}

Three clusters: {SAL, SUE}, {JOE, BOB}, {FRED}

Four clusters: {SAL, SUE}, {JOE}, {BOB}, {FRED}

These allocations seem not unreasonable when compared with the correspondence analysis ordination (Fig. 2.2), but which is best? A number of "stopping rules" have been developed for K-means and other cluster analyses, but unfortunately, the performance of any rule is data dependent (Milligan & Cooper 1987), and different rules are appropriate in different circumstances (e.g., Rendell & Whitehead 2003). Unless the clusters are very clear, I suggest that a range of numbers of clusters be tried, considered. and presented when using K-means.

Network analysts have come up with a number of methods of nonhierarchical clustering of individuals whose relationships are described by a similarity matrix (often just binary matrices consisting of 1s and 0s). A few of these methods can use nonbinary data such as a matrix of association indices and may be very useful (Section 5.7). For the social analyst, the network techniques have the advantages over K-means in that they work from similarity matrices rather than data matrices, suggest the optimal number of clusters (or the possibility that it is optimal not to divide the data), and provide a measure of the effectiveness of the division, modularity.

With any clustering, an important issue is whether the classification is appropriate in itself. If SAL and SUE are actually a permanent distinct social unit, then it probably is. If, on the other hand, they are just particularly "good friends" in a loose fission–fusion social system in which individuals associate with a range of others at different rates, then it probably is not. Instead, ordination methods (Fig. 2.3), as well as sociograms (Section 5.2), provide a more legitimate visualization of the society.

There is the same concern with the other major class of cluster analyses: hierarchical methods. In hierarchical cluster analysis, clusters are formed either by division or agglomeration. In divisive techniques, all

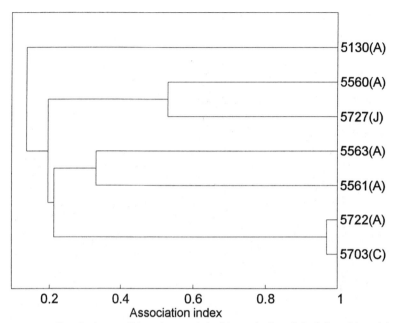

FIGURE 2.4 Example of average-linkage cluster analysis of the matrix of association indices of the social unit of sperm whales in table 2.5.

subjects start in the same cluster, which is then split into two (often by K-means), and these are then split, and so on, until each subject is in its own cluster. In contrast, in agglomerative cluster analysis, "distances" are calculated between all subjects to be clustered. These can be dissimilarities, or inverse similarities. Each subject starts out in a cluster by itself, then the two clusters with the smallest distance between them are joined, and then the next two, and so on, until all subjects are in the same cluster. There is a range of available agglomerative techniques that differ in how the distances between a newly formed cluster and the other clusters are calculated, including "single-linkage," "complete-linkage," "average-linkage," and "Ward's" techniques. Of these, average-linkage or Ward's technique is generally preferred (Milligan & Cooper 1987), perhaps especially in social analyses because extreme small or large distances, whether caused by random error, measurement error, or unusual individuals, have less impact on the results than when using single-linkage or complete linkage techniques (Whitehead & Dufault 1999).

The results of cluster analyses are usually displayed as a *dendrogram* or tree diagram (e.g., Fig. 2.4), which shows the pattern of cluster formation along the axis of the similarity measure. In Fig. 2.4, the sperm whale calf and its mother form a very tight cluster. Dendrograms are attractive: They seem to provide a model of a social system, but they are frequently overinterpreted and sometimes inappropriate (Section 5.2). If a society is hierarchical in its structure, with permanent or semipermanent units nested within other permanent semipermanent social entities ["tiers" (Wittemyer et al. 2005)], then hierarchical cluster analyses and the resultant dendrograms may be appropriate ways to visualize it. If no such hierarchical nesting of social entities exists in reality, however, then a dendrogram may be deeply misleading.

There is also a useful statistical check on the validity of a dendrogram, the cophenetic correlation coefficient, or CCC (Wittemyer et al. 2005). The CCC is the correlation between the input distance measure among all pairs of subjects and the level at which they are joined on the dendrogram (Bridge 1993). CCC's range from 1.0, a perfect fit, to 0.0, no relationship, and values greater than 0.8 are sometimes considered to indicate that the dendrogram is an acceptable representation of the input distances (Bridge 1993).

More information on classification and some examples are given in Sections 5.2 and 5.7.

2.8 Model Fitting and Selection: The Method of Likelihood and the Akaike Information Criterion

With data collected from a system, including a social system, we can fit mathematical models. In social analysis, a model might be that individuals form semipermanent units but move between units at a certain unknown rate. The models are probabilistic in the sense that they assign probabilities to certain data values. They usually contain parameters that we do not know, such as the mean unit size and rate of movement between units. The statistical term *likelihood* is simply the probability of obtaining a particular data set given a particular model and set of parameters. Intuitively, larger likelihoods indicate better model fits. Thus, a standard and theoretically appealing (Silvey 1975, pp. 68–79) manner of estimating the parameters of a particular model is to find those that maximize likelihood. Likelihood methods also allow the estimation of confidence intervals for parameters: The confidence interval includes sets of parameter values such that the likelihoods with these sets of values are within a particular factor (chosen depending on the confidence percentile required) of the maximum likelihood. An important book on likelihood is that of Edwards (1992).

In some cases, such as linear regression with normal errors, there are known analytical formulas for maximum likelihoods, and maximum likelihood estimators of parameters usually correspond to those in standard practice (minimizing the sum of the squares of the deviations of the differences between the actual data values and their predicted values). In other cases, there is no analytic formula for maximum likelihood, but likelihoods can be calculated for any set of parameter values using a computer, and the set of parameters that optimizes the likelihood can be found by optimization techniques. In still other cases, we need to use variants of the likelihood method, such as quasi-likelihood, which deals with overdispersed data (Wedderburn 1974), simulation, or summed log-likelihoods (Whitehead 2007).

Likelihood methods can also be used to choose between models. Thus, for each of two competing models, say, one including permanent units and the other not, we calculate and compare the maximum likelihoods over all possible parameter combinations. However, we cannot simply choose the model with the greatest maximum likelihood. If the models have different numbers of parameters, then the one with the most parameters will usually have the greatest likelihood, and if one is a special case of the other, then the general model will always fit better and have a higher maximum likelihood. In the interests of parsimony, and for other good reasons (Burnham & Anderson 2002, pp. 29–35), we choose the simpler model if the data fit it almost as well as its more complex elaboration.

The traditional manner of comparing the fit of data to two or more models is to use the likelihood-ratio test. The ratio of the maximum likelihood of the more complex model to that of the simpler one is logged, doubled, and compared to a chi-squared distribution with degrees of freedom equal to the additional number of parameters in the more complex model. If an unusually large likelihood ratio is found, then the null hypothesis that the simpler model is correct is rejected. The likelihood-ratio test has been much used but suffers from a number of problems. The most serious are that it is only possible to compare two models at a time, they must be nested within one another, the chi-squared distribution of the likelihood ratio is only an approximation except at very large sample sizes, and setting the problem up as a test for the "true" model, with one model being null and one the alternative, is unrealistic and misleading. Generally in the life sciences, no model is true; instead, we are looking for a model that best approximates a complex reality (Burnham & Anderson 2002, pp. 20–22). For these good reasons, the likelihood-ratio test has become less used, being replaced by the Akaike information criterion (AIC) and its variants.

Akaike (1973) proposed that "Kullback-Leibler information" be used as a fundamental basis for model selection. Kullback-Leibler information measures the discrepancy between data and model, and can be estimated from likelihood. If the maximum likelihood of the data set under a particular model is L and the number of estimable parameters is K, then

$$AIC = -2 \log_e(L) + 2K$$

This is a simple and very useful result. To compare the fit of a data set to several models, we calculate the AIC for each and select the model with the lowest AIC. The degree of support for the other models is indicated by the excess of their AIC over that of the best-fitting model, ΔAIC. For instance, if ΔAIC < 2 for a model, then it has some support and should not be dismissed. The shortcomings of the likelihood-ratio test are gone: Models that are not nested can be compared; there is a version of AIC (AIC_C) that compensates for small sample sizes; and, best of all, there is no "null model" or "true model." There is also a version of the AIC, QAIC, that uses quasi-likelihood and deals with overdispersed data, which may result from grouped animals behaving similarly. These and many other features of AIC are discussed by Burnham and Anderson (2002).

2.9 Computer Programs

To achieve realistic and useful social models, we almost always need a large amount of data (Section 3.11), and frequently we use permutation tests or other computationally intensive statistical methods, such as ordination and lagged association rates. Thus, computers have become essential tools in virtually all social analyses. Useful software is of four major types: that used during the collection of data; data storage and manipulation programs; general statistical analysis software; and specialized software for social analysis. There is overlap between types. For instance, database and spreadsheet programs, such as Microsoft's Access and Excel, although principally used for storing and organizing data, can be customized to collect data and can perform some statistical analyses. In the following summary and in Table 2.6, which lists data analysis software (together with web addresses), I describe some of what is available, noting those programs that are financially cost free. In some

Table 2.6 Software That May Be Useful for Social Analysis

Name	URL[a]	Free?	Notes
Data collection software			
JWatcher	http://www.jwatcher.ucla.edu	Yes	Input, storage and analysis of behavioral data
Noldus Observer	http://www.noldus.com	No	Collection, analysis, presentation, and management of observational data
Psion	http://www.psion.com	No	Hand-held data collection
Data storage software			
Excel	http://www.office.microsoft.com	No	Standard spreadsheet storage and analysis
Access	http://www.office.microsoft.com	No	Standard database software, linked to Excel
OpenOffice	http://www.openoffice.org	Yes	Free analogs of Excel and Access
General statistical analysis software			
Minitab	http://www.minitab.com	No	Easy-to-use statistical analysis
SPSS	http://www.spss.com	No	Easy-to-use statistical analysis
Systat	http://www.systat.com	No	Intermediate statistical analysis
SAS	http://www.sas.com	No	Intermediate statistical analysis
S-PLUS	http://www.insightful.com/products/splus/default.asp	No[b]	Powerful statistical analysis
R	http://cran.r-project.org	Yes	Powerful statistical analysis
Matlab (plus Statistics toolbox)	http://www.mathworks.com	No	Powerful and flexible general analysis (including sparse matrices)
PopTools	http://www.cse.csiro.au/poptools	Yes	Matrix manipulation, likelihood, simulations, bootstrap
Social analysis software			
MatMan	http://www.noldus.com/site/doc200401030	No	Manipulates matrices, good for analyses of dominance and reciprocity
SOCPROG	http://myweb.dal.ca/~hwhitehe/social.htm	Yes[c]	Wide range of social analyses
PeckOrder 1.03	http://www.animalbehavior.org/Resources/CSASAB/#PeckOrder_1.03.hqx	Yes	Dominance hierarchies, older, Macintosh based
Network analysis software			
	See http://www.insna.org/INSNA/soft_inf.html for a list		
UCINET	http://www.analytictech.com/ucinet/ucinet.htm	No[b]	Range of network and other analyses
Pajek	http://vlado.fmf.uni-lj.si/pub/networks/pajek	Yes	Range of network analyses
NetDraw	http://www.analytictech.com/netdraw/netdraw.htm	Yes	Visualizes networks
GraphViz	http://www.graphviz.org	Yes	Visualizes networks
Other			
KINSHIP	http://www.gsoftnet.us/GSoft.html	Yes	Estimates genetic relatedness
CAIC	http://www.bio.ic.ac.uk/evolve/software/caic/	Yes	Performs comparative analysis using independent contrasts, Macintosh based

[a]Subject to change over time.
[b]Free evaluation version available.
[c]To run SOCPROG in its original form, you need to have installed MATLAB plus the Statistics Toolbox, which are not free. However, there is a compiled version of SOCPROG available for which MATLAB is not necessary.

cases, however, I have not tried the software and rely on the descriptions of the authors and vendors, which may or may not be accurate. In Appendix 9.3, the capabilities of the different analytical packages are indexed to the sections of the book that describe methods for which the packages are useful.

Software is in flux. The information in the following subsections, Table 2.6, and Appendix 9.3 is accurate and fairly complete at the time of writing. However, new methods will be incorporated into packages, less-than-useful procedures will be dropped, new packages will be introduced, manuals will be rewritten, and web addresses will change. Thus, I cannot guarantee that the information presented will all be accurate and relevant at the time you are reading this. You may need to explore.

2.9.1: Data Collection Software. There are a number of software packages that can assist considerably in the collection and storage of behavioral data. Use of such packages can reduce or remove the need for written field notes, data sheets, and voice records, all of which can be cumbersome, time-consuming to transcribe, and sources of error. Commonly used is the Noldus Observer, described by the provider as a "system for the collection, analysis, presentation and management of observational data"; it can be used "to record activities, postures, movements, positions, facial expressions, social interactions or any other aspect of human or animal behavior." There are versions for field and laboratory settings and the analysis of video data.

A comprehensive free package designed for the input, storage and analysis of behavioral data is JWatcher (Blumstein and Daniel 2007). As it is written in Java, it works on almost any type of microcomputer. There is also JWatcher-Palm that can be used to acquire data on a Palm equipped device.

The versatility of standard database, spreadsheet, and other programs, such as Access and Excel, allow scientists easily to design their own data input routines, and this is often done. In field situations, small "palmtops," such as those manufactured by Psion and Palm, can be programmed easily to collect systematic social data. Lehner (1998, pp. 162–266) describes older data collection software.

In some situations, the major challenge is not collecting the behavioral data, but identifying the individuals. Computers can be of assistance here, for instance. in processing photoidentifications (e.g., Araabi et al. 2000).

2.9.2: Data Storage Software. In many cases, data are manually transcribed from video or audio recordings or data sheets to databases or spreadsheets. It is usually both desirable and easy to export data collected on palmtops or produced by data collection software, such as the Noldus Observer, or automated or semiautomated data-processing systems such as Passive Integrated Transponder (PIT)-tag readers or photoidentification programs into databases or spreadsheets. Of these, Microsoft's Access and Excel are most popular. OpenOffice provides similar, but free, packages. Useful formats for data coding are discussed in Section 3.8. Within such packages the data can be manipulated in a large range of ways. Some analyses can be performed within Excel. The free PopTools Excel add-in is useful for social analyses. Data stored in Access or Excel can be input into the powerful general statistical packages and often, sometimes with a little manipulation, the more specialized social analysis software (see later discussion). Relational databases, such as Access, are particularly useful for storing information from multifaceted studies on the same individuals.

It is often useful to think about the data analyses to come when setting up data storage software. Some guidelines are given in Section 3.8.

2.9.3: General Statistical Software. There are several widely used packages for the statistical analysis of data that can be used to perform some of the social analyses in this book, such as ordinations and cluster analyses. Two of the easiest to use are Minitab and the Statistical Package for the Social Sciences (SPSS). Somewhat more sophisticated, but still simple to use, are Systat and SAS. Professional statisticians often use S-PLUS or the similar, but free, R. These are both very powerful and flexible packages but are more difficult to use than Minitab, SPSS, Systat, or SAS. An alternative for those with some programming skills is the extremely general and powerful programming package Matlab with its Statistics Toolbox. Matlab is particularly useful with large data sets because of its ability to use sparse matrices. Sparse matrices have a large proportion of zeros and are frequently encountered in social analysis of large populations because most pairs of individuals have not been observed interacting or associating. Special programming methods in Matlab can easily and efficiently store sparse matrices and thus allow analyses of large data sets that would not be possible otherwise.

2.9.4: Specialized Software. Many important social analyses, such as permutation tests (Section 4.9), cannot be carried out using the standard statistical packages without additional programming (usually easiest in S-PLUS, R, or Matlab). Therefore, several packages of software have been developed.

Noldus MatMan is designed for matrix manipulation and analysis. According to the provider, it is "a powerful tool for performing a wide variety of ethological analyses on sociometric, behavioral profile, and behavioral transition matrices." It is an easy-to-use add-in for Excel, and is particularly useful for the analysis of dominance hierarchies (Section 5.4) and reciprocity (Section 4.8).

My package SOCPROG is a set of Matlab (plus Statistics Toolbox) programs for social analyses (as well as some movement and population analyses). It can be used as it stands or be modified by anyone familiar with programming in Matlab. SOCPROG is free. To have the ability to modify programs, the user needs access to Matlab and its Statistics Toolbox, which are not free. There is also a stand-alone compiled version of SOCPROG for which Matlab is not required. SOCPROG is particularly useful for manipulating large identification data sets (changing sampling periods, definitions of association, restricting data), for ordinations (Section 5.2), permutation tests (Section 4.9), lagged association rates (Section 5.5), and multivariate methods (Section 5.6). It also contains modules for population and movement analyses and utilities for exporting the data in the format used by other programs, including those used for network analysis.

PeckOrder 1.03 is an older Macintosh program for analyzing dominance hierarchies.

JWatcher, in addition to aiding the collection of behavioral data, performs a number of analyses. These include calculating time budgets and the analysis of sequences (Blumstein and Daniel 2007).

Network analysis is an increasingly used set of techniques in the analysis of animal societies (Section 5.3). There are a number of programs that can carry out network analyses (see http://www.insna.org/INSNA/soft_inf.html for a list), including the following:

- UCINET (Borgatti et al. 1999) is a Windows-based program that performs a range of network and other analyses (and includes Pajek and NetDraw; see following items). There is a free evaluation version.
- Pajek is another Windows-based program, and is free for noncommercial users.
- NetDraw visualizes networks using UCINET files. It is free.
- GraphViz is a free set of programs that is very useful for representing and analyzing very large networks.

Genetic relatedness is becoming an increasingly important element of social analyses (Sections 4.2 and 7.3). The genetic relatedness among dyads is most often estimated by molecular genetic techniques, particularly using microsatellites (Selkoe & Toonen 2006). There are several programs for processing such data, but Queller and Goodnight's (1989) KINSHIP is a general favorite for producing relatedness estimates.

The method of independent contrasts has become very important when making comparisons between species or other taxonomic levels (Box 7.1). The program CAIC is frequently used for this (Purvis & Rambaut 1995).

3 Observing Interactions and Associations: Collecting Data

3.1 Types of Behavior

Having decided what we wish to study and why (Chapter 1), the next step in analyzing social structure, and the subject of this chapter, is the collection of data. The real-life behavior of individual animals needs to be abstracted into a form, usually a numerical form, in which it can be analyzed. So, how should behavior be described? This subject is considered in detail by several authors, for instance, Lehner (1998, pp. 109–124) and Martin and Bateson (2007, pp. 48–61), and so in this chapter I summarize areas well covered in these books, concentrating on the most significant issues for social analysis.

There are two general ways of describing behavior—in terms of either the structure of the behavior (physical form, posture, movements, etc.) or the consequences of the structure for the animal or the environment (Martin & Bateson 2007, p. 32; Lehner 1998, p. 81).[1] Consequences might include agonism, feeding, or grooming. Describing behavior in terms of consequences is generally more powerful and economical (Martin & Bateson 2007, p. 32), but sometimes when we do not understand the behavior of

1. Lehner calls these types "empirical" and "functional" descriptions, respectively.

the animals well, it might be better to stay with structural definitions of behavior types. An example is when cetaceans are viewed from the surface and a characteristic posture of infants seems to suggest suckling, but nipple contact cannot be observed and we are not certain about the function of the behavior (Gero & Whitehead 2007).

Another important distinction is between behavioral events and behavioral states (Altmann 1974). Events are virtually instantaneous, and one must watch for awhile to see any. In any time period, they can be counted and classified. Events include behavior such as vocalizations, sudden movements, and ingestion of prey. States are behavior patterns of relatively long duration. Often, several states are defined [e.g., foraging, traveling, resting and socializing (Mann 2000)] such that an animal is always in at least one of them. They may be defined as exclusive, so that an animal is only in one state at a time, or be allowed to overlap (e.g., social and traveling). Events occur within states, and can be used to define them: The state "feeding" might consist of time periods that include events of food ingestion. Before starting a study, events and states should be defined as rigorously as possible, perhaps with the help of a preliminary study whose data are not used in the final analysis (Martin & Bateson 2007, pp. 31–32). If behavior is videotaped or recorded acoustically, then some of these decisions can be postponed, but rigor is important.

In this book, I am concerned with social structure, and therefore social behavior, which involves two or more individuals. The fundamental element of social behavior is the interaction (Section 1.6), an event that can be defined structurally or, more usually, by consequence. However, we can also use interactive state measures, associations, as the core of our social analysis or combine interactions and associations. Interactions and associations are the subjects of the next two sections, which are followed by a section on groups. Groups can be used as shortcuts to the designation of associations. All of this depends crucially on the ability to identify individuals (Section 3.5). Thus, the meat of most analyses of social structure comprises systematically collected records of observations of interactions, associations, or group membership among identified individuals. These may be supplemented by records of rare but important behavior and supplementary data about the individuals such as sex, age, reproductive status, or kinship (Section 3.6). The choice of sampling scheme depends on whether we are collecting interaction or association data, and other factors. These are considered in Section 3.7. The final sections of this chapter consider the formatting, structuring, and power of data sets.

3.2 Interactions

Interactions are the basis of Hinde's (1976) framework for the study of social structure. The definition of interaction is fairly straightforward ("the behavior of one animal is affected by the presence or behavior of another"), but the operational use of the term is less so (Hinde 1976). Is a fight one interaction, or are all the elements (physical contacts, vocalizations, movements, wounds, etc.) individual interactions? Do we separate different types of affiliative behavior or simply have one interaction type that may include affiliative grooming and vocalizations as well as movements?

My advice is to collect data at the highest level of resolution that permits operational consistency, so that interaction types are rarely missed or misclassified. If high operational resolution is feasible, it may turn out that some distinctly defined interaction types are functionally equivalent to the animals (interaction types A and B are interchangeable from their perspective) or redundant (A always follows B). A well-organized analysis will identify such relationships between interaction types and account for their interdependence in subsequent analyses (e.g., using principal components analysis; Section 2.6). Nothing is lost by collecting the high-resolution data, even if they are somewhat redundant. More frequently, especially with hard-to-study species, achievable operational precision will be much less than that which is meaningful to the animals. In these cases, the more detailed the data, the more meaningful is the model of social structure that results, provided that the data are reliable. For some analyses, however, it may be efficient to lump types of interaction into classes based on structure or function, such as "vocalizations" or "agonistic behavior." Although high-resolution data can be lumped if desired, it is not possible to go the other way, increasing resolution.

Sometimes, interactions are identified by synchrony or leader/follower events—one animal follows another in performing a particular activity— in other words, by temporal patterning (e.g., Connor et al. 2006). To identify synchrony or leader/follower events rigorously, we need simultaneous records of behavioral events from several individuals.

To give an idea of the range of types of interactions that can be recorded, Table 3.1 lists some, noting whether they are defined by structure or consequence (Section 3.1) and whether they are *symmetric* (if A interacts with B, then B interacts with A).

In what follows, I usually assume that interactions are dyadic—involving two animals—although some of the analyses make sense with triadic or higher-order interactions (e.g., Kummer et al. 1974).

Table 3.1 Examples of Interaction Types That Can Be Used as the Basis of a Social
Analysis Together with Whether They Are Primarily Structural or a Consequence, or
Primarily Symmetric or Asymmetric

Type	Structural/consequence	Symmetric?
Grooming bout	Consequence	No
Fight outcome	Consequence	No
Touch	Consequence	Yes
Synchronous dive	Consequence	Yes
Leader/follow dive	Consequence	No
Vocalization exchange	Consequence	Yes
Suckle	Structural	No
Intromission	Structural	No
Mating	Consequence	Yes
Particular gesture in dyadic context	Structural	No

3.3 Associations

Behavioral interactions, the foundation of Hinde's conceptual framework of social analysis, are events. Interactions cannot always be observed, however, and in some more cryptic species are virtually always
hidden. A common way around this difficulty is to use *associations* instead of interactions or in addition to interactions as the fundamental
elements of social analysis. Dyads are in "association" if they are in a
situation in which interactions usually take place (Whitehead & Dufault 1999). Associations are state measures, and usually they are more
easily measured than interactions. They can often be determined from
nearly instantaneous observations, whereas interactions, even when observable, require prolonged observation.

Association can also be reasonably interpreted as "within range of
communication" because communication involves the active or passive
transmission of information that may change the behavior of the recipient (Bradbury & Vehrencamp 1998, p. 2), resulting in an interaction.
This emphasizes the important role of communication in studies of social structure (Costa & Fitzgerald 1996).

Thus, ideally, the social analyst would initially make a thorough
study of communication in her study species or have access to the results
of one. From this, she could determine the dyadic circumstances that
best characterize communication between a pair of animals. She can then
define association in such a way that it delineates circumstances under
which communication, and interactions, take place. Systematic records
of such associations are then used as the data for social analysis. Such
rigorous approaches are rare. Instead, even the best studies rarely go
beyond reasoning such as the following: Animals can hear each other at

Table 3.2 Examples of Definitions of Dyadic Association That Can Be Used as the Basis of a Social Analysis Together with Whether They are Symmetric or Asymmetric

Definition	Symmetric?
Within *x* body lengths (or *x* meters)	Yes
Within *x* body lengths (or *x* meters), and in same behavioral state	Yes
Within *x* body lengths (or *x* meters), and heading in same direction	Yes
Nearest neighbor	No
Sharing feeding site/nest/roost	Yes
Duetting	Yes
Grooming	No
Overlapping home ranges	Yes
Grouped (table 3.3)	Yes

ranges up to about x meters, and so we will define association as "dyads separated by less than x meters." This is not unreasonable, and in most cases, errors in choosing a suitable x will not profoundly affect the subsequent analysis. A too small value of x will omit some interactions, and a too large one will include noninteracting dyads, but if a large data set is collected and there is no systematic bias (such as might be caused by pairs of individuals who generally interact at ranges just greater than the chosen x), an informative social model should emerge.

Association is usually defined based on spatial proximity plus, perhaps, some behavioral state measure (e.g., "within x body lengths and heading the same direction"). It is often possible, and desirable, to measure more than one association measure simultaneously. Perry (1996), for example, noted associations within 1, 5, and 10 body lengths for capuchin monkeys (*Cebus capucinus*); animals within 10 body lengths may interact vocally, and those within 1 body length may interact using touch. Table 3.2 lists some ways in which association has been defined.

Associations may be asymmetric. For instance, nearest neighbor is a commonly used asymmetric definition of association. During an observation, A may be the nearest neighbor of B, but B is not necessarily the nearest neighbor of A. Most analytical techniques assume symmetric association measures, so I generally recommend their use, although it should not take too much work to adapt them for asymmetric measures.

Sometimes, the same behavior can be viewed as either an interaction (event) or an association (state). The difference is whether the observations are considered continuous or instantaneous. For instance, we can count the number of grooming bouts in an observation period, considering each bout as an interaction, or observe whether a dyad is engaged in grooming behavior when we observe them, in which case grooming is an asymmetric association measure.

Association within any sampling period (Section 3.9) is usually defined in a 1:0 manner; a dyad is observed associated or is not observed associated within that sampling period. It is possible to define association as a continuous variable [e.g., a function of proximity or dive synchrony within the sampling period (Whitehead & Arnbom 1987; Perry 1996)], but this adds considerably to the complexity of the subsequent analysis, and I do not usually recommend this.

3.4 Groups

A simplifying assumption that is often made is that all individuals within some spatiotemporal "group" are associated with each other. Such groups are assumed to be *transitive* in the sense that if, at any time, A and B are members of the same group and so associated, and B and C are members of the same group and so associated, then A and C are also members of the same group and associated. We have called the assumption that grouped individuals are associated "the gambit of the group" (Whitehead & Dufault 1999). The fundamental assumptions, almost never tested, are that all, or almost all, interactions of some type take place within groups and that interactions of this type are similar and occur at a similar rate among all animals within a group (Whitehead & Dufault 1999).

To evaluate the likely validity of the gambit of the group, it helps to consider the reasons that animals may be in spatiotemporal proximity. An important distinction is between spatiotemporal clusters of individuals that are entirely the result of some nonsocial forcing factor, such as a localized source of food or shelter, and those that result from the active behavior of individuals converging on, or maintaining proximity with, other animals. I call these *aggregations* and groups, respectively, and only groups are of direct interest as elements of social structure. Connor (2000), making the assumption that animals do not usually behave maladaptively, refers to them as nonmutualistic groups (= aggregations) and mutualistic groups (= groups) because individuals are likely to seek or maintain proximity with other individuals if and only if there is expected to be mutual benefit. This suggests two ways of distinguishing aggregations from groups. They are groups if it can be shown either that individuals actively seek or maintain proximity with other individuals or that there is some benefit of being grouped with others.

In some cases, these criteria allow the simple recognition of groups. For instance, individuals clustered over a habitat with uniform resources (flocks of roosting birds in some trees when other nearby and similar trees are empty, or ungulates migrating over featureless habitat in groups) can be considered groups, as may clusters of individuals that are passed over by a predator that takes lone individuals. In other cases, it is not so clear whether clustered individuals form a group. Bats may be clustered in a cave solely because it is suitable roosting habitat or because they are drawn to assemblages of other roosting bats. Of course, even within my definition of group there is enormous variability. A fish school may contain animals drawn together solely to combat predation and possess little temporal stability and no behavioral substructure. In contrast, a nearby pod of killer whales (*Orcinus orca*) may be made up of genetic relatives who spend their lives together, feed together, and have distinctive relationships with each other (Ford et al. 2000). Such distinctions are considered in the following chapters, but at this step, we need to exclude aggregations from further social analysis.

Groups are usually obvious. If the spatiotemporal clustering is so subtle that human observers cannot be sure that it is present or so variable in type that they find it difficult to come up with a rigorous definition, then they should doubt whether such groups are meaningful to the animals. In many cases, however, although group distinctions are generally obvious, there are some borderline cases ("Is that one group or two?"). It is important that such instances be treated consistently, and so we need a criterion for allocating individuals to groups (Martin & Bateson 2007, pp. 46–47).

Ideally, we would base our definition of group on studies of communication, as with association (Section 3.3), but this is rarely done explicitly. In some fortunate instances, groups consist of sets of animals using small areas or volumes of suitable habitat separated clearly from other such groups. Roosting sites of birds and bats in particular trees, leaves, or caves are such cases (e.g., Vonhof et al. 2004), as are islets used by seals to haul out (e.g., da Silva & Terhune 1988). More usually, groups are formed over homogeneous or continuously varying habitat. A useful empirical approach is to measure interindividual distances, perhaps on photographs or video, and examine their distribution. If clear modes are apparent, then these can suggest a suitable definition of a group (Clutton-Brock et al. 1982, pp. 319–320). For instance, if there are many interindividual distances between 1 and 5 body lengths but very few between 6 and 30, then perhaps animals within 5 body lengths should be considered grouped. Whereas such a distance criterion works well with associations (Table 3.2), however, it may give inconsistent results when used directly to define groups: A and B may meet the criterion, and so may B and C, but A and C may not. In this case, the transitive feature of groups is violated. Frequently, researchers use a

Table 3.3 Examples of Criteria Used to Distinguish Groups

Definition
Sharing feeding site/nest/roost/haulout
Within x body lengths (or x meters) chain rule
Within x body lengths (or x meters) chain rule, and in same behavioral state
Within x body lengths (or x meters) chain rule, and coordinating movement
Clusters produced by kth nearest-neighbor clustering on spatial arrangement of individuals (Strauss 2001)

"chain rule" to circumvent this problem (Clutton-Brock et al. 1982, pp. 319–320; Smolker et al. 1992). If A and B meet the criterion as well as B and C, then A and C automatically meet it too, and so we have transitive groups.

Strauss (2001) considers the difficult case of fluid shoals of fish, which may form irregularly shaped groups. After some experimentation, he found that a method called "kth nearest-neighbor hierarchical clustering" best mimicked human perceptions of shoal membership. He also developed permutation methods to test whether the spatial arrangement of the animals is clustered compared with a random null hypothesis.

Table 3.3 list some criteria used to designate groups. Group criteria may include a behavioral condition such as "coordinated movement" or "in same behavioral state," which helps to exclude individuals that are incidentally clustered (Mann 2000). As with associations, it is possible simultaneously to use two or more types of group at different spatial or temporal scales, usually with one type nested within the other, which thereby stand in for different classes of interaction.

Estimating group size is usually easy if individuals can be assigned to groups: We just count. In some cases, however, for instance, with cryptic animals, a population assessment technique may be useful for estimating group size. Such techniques are summarized in Appendix 9.5.

3.4.1: Typical Group Size. A distinction that is important in many cases is between mean group size and mean *typical group size* (Jarman 1974). The former is the mean size as experienced by an outside observer, such as a predator or the social analyst, the latter is the size as experienced by a member of the population. For example, if there are four groups of size 1, 2, 2, and 3, then the mean group size is 2, whereas the mean typical group size is 2.25. There is 1 animal in a group of size 1, 4 in groups of size 2, and 3 in groups of size 3, giving a mean typical group size of $(1 \times 1 + 4 \times 2 + 3 \times 3)/8 = 2.25$. Mean typical group sizes are usually higher, and never lower, than those experienced by outside observers.

If we have counts or estimates of the sizes of n observed groups $N_g(i = i, \ldots, M)$ their mean is simply $\Sigma N_g(i)/M$, whereas the mean typical group size is $\Sigma N_g(i)^2/\Sigma N_g(i)$. We can also speak of the mean typical sizes of other social or nonsocial entities, such as aggregations, units, or communities. As an example of the differences between an animal- and an observer-centered approach, here are the mean sizes and mean typical sizes of aggregations (probably a nonsocial assemblage), groups, and social units (both social assemblages) of sperm whales (*Physeter macrocephalus*) (Whitehead 2003, pp. 213, 218):

	Mean	Typical mean
Aggregation (Ecuador 1991)	60.1	77.4
Group (South Pacific)	19.4	25.1
Unit (Galápagos)	10.5	13.6

3.5 Identifying Individuals

A primary requirement for the social analyses described in the subsequent chapters is that individuals be identifiable. There are many ways to do this. With humans and some small populations of other species, it may be possible to discriminate visually all individuals reliably in real time. In larger populations, photographs of natural markings can be used either to ground-truth visual identifications or as the sole source of identifications (Pennycuick 1978; International Whaling Commission 1990; Lehner 1998, pp. 221–223). Vision is not the only medium, however, by which individuals can be distinguished. In appropriate circumstances, individual identification may be possible using vocalizations (Adi et al. 2004) or by collecting DNA samples (Palsbøll et al. 1997; Sloane et al. 2000). Artificial marks, including dye marks and tags, can be used to identify individuals visually (Stonehouse 1978), acoustically (Zeller 1999), or through radio signals (Chambers et al. 2000). Use of passive integrated transponder (PIT) tags that can be implanted into animals or attached to them and "read" by an external radio signal is becoming an increasingly important method of identifying animals (Biomark, Inc., see http://www.biomark.com/; e.g., McCormick & Smith 2004).

In many situations, especially if tags or artificial marks are used, but also sometimes with naturally marked animals (e.g., Ottensmeyer & Whitehead 2003), only a portion of the population will be identifiable. This should be considered in subsequent social analyses. For instance, estimates of group sizes need to be corrected for animals that are not identified. In the worst scenario, the social behavior of the identified animals is not representative of the population. This could happen if more-aggressive individuals were more likely to accumulate natural markings or those generally in smaller groups were more likely to be tagged.

Another important issue with individual identification is reliability. With almost all techniques, there is a chance that an animal can lose or change the identifying feature. Natural marks can change, tags may be lost, or the technology in acoustic or radio tags may malfunction. In most cases, a numerical analysis treats this as equivalent to mortality or permanent emigration. Thus, if marking failure is likely, then methods and models that include the disappearance of the animal from the population, by whatever means, need to be used. If "mortality" or "survival" estimates are produced (as in some models of lagged association rates; Section 5.5), then it must be recognized that these include mark failure, as well as perhaps permanent emigration from the population. With natural markings, there is the additional possibility that an unrecognizable mark change will produce a "new" individual in the population, for instance, when a large, new mark obliterates the features previously associated with an individual or a "clean" unmarked individual gains marks and joins the study population. Although this is an important concern when estimating populations using natural markings (Hammond 1986), these changes, as long as they are not too frequent, are unlikely seriously to affect social analyses.

Misidentification is of greater concern. This can occur with all techniques and at several stages of the identification process. Real-time visual identifications can be wrong, acoustic identifications may be less than 100% accurate (Adi et al. 2004), and equipment problems can cause errors with acoustic or radio tags. The data can be recorded wrongly, entered in the database erroneously, or scrambled by a computer. All of these can cause data to enter the numerical analyses tagged with the wrong individual. With large data sets, there are almost certain to be some errors. One hopes that the rate will be small, but what will be the effect on the output measures of social structure? I cannot give an overall prescription, and each situation should be considered on its merits. The greatest danger, however, occurs when a particular data record can have a considerable effect on the output. This might be the case if we are looking for closed units, when a misidentification could mean rejecting this hypothesis or lumping separate units. With less hard-and-fast social structures, however, such as "fission–fusion societies," a small proportion of misidentifications is unlikely to have any major impact on the results. One can test the effects of misidentifications by deliberately making some and seeing how the output is changed. This method will not work, however, if errors have already ruled out a "true" model for the data, such as closed units.

3.6 Class Data

An important part of Hinde's (1976) framework (Fig. 1.4) is the generalization from interactions between two individuals to interactions between classes of individuals. This is often done by some form of averaging ("the mean rate at which mothers groom their neonates") or by abstracting essential characteristics of interrelationships between classes ("fights only occur between mature males") (Hinde 1976). These class-abstracted results are clearly of interest in their own right. When data are few or sparse, however, there may be insufficient power to categorize social structure at the level of individuals, and so class abstractions become the principal results of a social analysis (Section 4.11). Classes can also be used to form nonsocial measures of relationship (e.g., "same or different gender"; Section 4.2) that are important when trying to address functional questions (Chapter 7).

Thus, social analyses are much richer if animals can be classified using attributes of individuals (Table 1.2). Here are some of the classes most frequently used:

- · *Sex.* Gender can be determined by observation, photographs or video of genital areas or sexually dimorphic anatomical features; observations of gender-specific behavior, such as nursing; or sex-specific DNA markers in tissue samples.
- · *Age.* Ideally, this is available from the lifetime knowledge of an individual, but sometimes accurate aging of living animals can be achieved through other means, such as drawing and sectioning a tooth. Age can often be estimated by size. With inaccurate aging methods, it may be more appropriate to assign animals to general age classes.
- · *Physiological state.* Classes may describe sexually mature or immature animals, pregnant or estrous animals, or some other physiological state.
- · *Subspecies, morph, and so on.* These can be considered in mixed populations.
- · *Matriline* (or *patriline*). These may be used in populations with well-known genealogies.

- *Genes.* The genetic class that is most usually employed is the mitochondrial haplotype (e.g., Weinrich et al. 2006), which in some respects stands in for the matriline when this is not known.
- *Behavioral phenotype.* Individuals can be classified into those that are dominant, aggressive, submissive, and so on (see Section 4.3).
- *Social unit, community.* Sometimes, the results of one level of social analysis can be used to define classes (Section 5.7) and these used to investigate questions such as whether patterns of within-unit or within-community social structure differ among units or communities or whether there are consistent affiliations among units.

3.7 Collecting Social Data

The analysis of social structure needs data. Social data may be recorded by human observers on data sheets, voice recordings, or photographs or keyed straight into computers (Section 2.9). Alternatively or additionally, acoustic, visual, or electronic data may be recorded by automated devices.

From the perspective of Hinde's framework and this book, these data are in the form of records of interactions and/or associations among identified individuals. Suppose a researcher is interested in a population of animals some or all of which are identifiable, and that she has a time frame over which the study is to be carried out and an effort budget. How should she plan data collection to give the most informative model of social structure? This is not an easy decision. It depends on the actual social structure, what behavior is observable, and a number of other factors.

Altmann's (1974) paper "Observational study of behavior: sampling methods" has been very influential in guiding the collection of behavioral data and has formed the basis of several other good reviews of protocols and procedures in behavioral observation (e.g., Martin & Bateson 2007, pp. 48–61; Lehner 1998, pp. 189–210; Mann 2000). Social data form a subset of behavioral data, and a subset with special characteristics, because two (or possibly more) individuals are involved. Thus, Altmann's (1974) recommendations need some refinement. In the following subsections, I consider whether to collect interaction, association, or group data and how to collect them. Table 3.4 summarizes the major recommendations and contrasts the features of interaction and association data.

Table 3.4 Collecting Interaction or Association Data: Guidelines

	Interactions	Associations
Type of measure	Event	State
Dyadic measure	Usually counts	Usually 1:0 (associated:not associated)
Use when	Interactions reliably and frequently observable	Interactions not reliably or frequently observable; coordinated behavior predominates
Follow protocol	Usually individual follow is best	Usually survey or group follow is best
Sampling protocol	All interactions involving focal animal with times and interactant identities is best	Associations are noted at regular times or when they change

3.7.1: Interactions, Associations, or Groups? As indicated in preceding sections, social analysis can be based on interactions, association, or groups. But which is preferable?

If interactions form the basis of social structure and associations are merely the imperfectly defined circumstances under which interactions are likely to occur (Section 3.3), then would it not be true that interactions should be preferred over associations as the targets of data collection? Under this rationale, it should be better to record when animals touch than that they are grouped, and associations should only be used as a "stand in" when interactions are unobservable. This used to be my perspective (Whitehead & Dufault 1999). After more reflection, however, I am not so sure (Whitehead 2004). A pair may have an important relationship but not touch or perform any overt interaction. A seamless behavioral synchrony without any observable interactions, as is sometimes characteristic of dolphins (Connor et al. 2006), might indicate the strongest of relationships. From the practical perspective, if we can rarely see animals interacting, associations will be more appropriate measures of sociality. In such circumstances, records of associations may be much more revealing than those of interactions.

In many circumstances, either associations or groups can be recorded. To decide which is preferable, we would ideally need to see inside the minds of the animals. Are their locations, movements, and behaviors more the result of the locations, movements, and behaviors of particular companions or of "the group" itself? Resolving this would often be a major study in its own right, but aspects of the animals' behavior can help. When movement is coordinated within a whole group or there are frequent changes of position within the group such as occurs in some fish schools, then the group appears to be a more significant behavioral determinant than the identity of any associate. In contrast, if behaviors and movements vary among clustered animals or they show little active behavior or movement, as in resting lions (*Panthera leo*), then perhaps associations should be used. If accurate positions are recorded, then both are possible, associations being defined by, for instance, one of the measures in Table 3.2 and then groups formed from the associations using a "chain rule" (Section 3.4). Group memberships are usually easier to record than associations, however, and so practical limitations may play a role in this decision.

There is no theoretical problem with recording interactions as well as associations and perhaps also groups (potentially derived from the association data), and more than one type of association, interaction, or group. If the different dyadic measures are well correlated and add little information, then methods such as principal components analysis can be used to simplify the multivariate data set (Section 5.6). If they are not well correlated, then the analysis is that much the richer.

3.7.2: Temporal Patterning and Length of Observations. Let us suppose that we have start and stop dates for the research, say, 1 June to 10 September, and also a limit on the total amount of research effort, say, 100 hours of observation. How should it be allocated? There are many possibilities, such as 1 hour per day every day for 100 days, or ten 10-hour days on 1 to 5 June and 26 to 31 August.

To make the best use of time resources, we need to have some idea of the temporal patterning of social relationships. At one extreme, if groups of bats are defined on the basis of roosting in a cave together and no bats change caves during the day, then there is no point in spending all day watching the bats in one cave. Instead, just enter the cave for long enough to identify the animals using it, move to the next cave and identify its inhabitants, and end the day's observation when all caves, or some predetermined proportion of them, have been sampled. At the other extreme, if fish are continually changing their associations, longer periods spent with any individual or group or in any area may be appropriate.

Clearly, to study social structure over any time scale means that we need data over that scale. Thus, if the important social time scales are unknown, it makes sense to arrange the data collection so that a range of scales can be examined. Thus, if we collect data for 6 hours per day on days 1, 2, 3, 11, 12, 13 June; 1, 2, 3, 11, 12, 13 July; and 1, 2, 3, 11, 12 August, we can examine scales of up to 5 hours, 1 to 3 days, about 10 days, 30 days, and 60 days within a total of roughly 100 hours of observation, a quite diverse set of spans. It is also important to consider the timing of important activities, such as breeding periods, that may affect interactions, associations, and social structure.

3.7.3: To Follow? What to Follow: Individuals or Groups? Another important decision when planning studies of social structure is who to watch and for how long. An initial choice is whether to follow or survey. In a *survey*, an individual or group is first encountered and then observed, and then the researcher, or her eyes or binoculars, move on to another individual or group. In a *follow*, the researcher's attention stays with an individual or group. The survey-or-follow decision is sometimes trivial, but at other times, it is more challenging.

If one is studying a small, captive, low-energy population in an open habitat, then perhaps all individuals and each action can be seen. In such cases, it is possible and optimal to follow everyone, recording interactions and/or noting changes in association and groups. At the other extreme are large, active populations most of whose habitat is invisible. Dolphins of the open ocean are an example. We cannot consistently follow either individuals or groups or observe interactions. We are constrained to survey individuals as they are encountered and to note associations or groups. In intermediate situations, a range of factors comes into play when choosing an observational strategy.

To record interactions (events), we must follow, at least for short periods, because, by definition, no interaction is visible in an instantaneous survey. If associations (states) are the measures of choice or necessity, however, then surveys will be generally more efficient if the rate at which individuals change associations is less than the rate at which new groups of animals can be surveyed (Whitehead 2004). For example, suppose group composition changes about once an hour; then, in terms of producing a model of social structure from records of group membership, it is more efficient to leave each group after noting its membership as long as another group can probably be found within 1 hour.

If the decision is made to follow, then should individuals or groups be the subject? Obviously, the more animals on which data can be collected simultaneously, the more powerful is the analysis. If the group is small and interactions are infrequent and easily seen, then group follows in which all interactions among all individuals are recorded are optimal. More normally, however, if interactions are the social measures of choice, then these will be difficult to record systematically for a whole group (Altmann 1974). Thus, if interactions are being recorded, then usually they should be between a focal individual that is being followed and others. In particular circumstances (such as parent–offspring or courting pair), dyadic follows may be appropriate, with interactions between the focal pair and between the focal pair and others all being recorded. Although recording all interactions within a group is usually impracticable (Altmann 1974), however, it is often possible to record all associations, especially if using the "gambit of the group" so that the group itself is used to define associations. In such cases, group follows will provide more information per unit time.

It is sometimes possible, and profitable, to make hybrid follows. For instance, while tracking a group of sperm whales (*Physeter macrocephalus*) for periods of days, we may carry out focal-individual follows of individuals during the 8 to 10 minutes that animals spend at the surface between dives (Whitehead 2004). Similarly, when following a large group of ungulates, one can make surveys of subgroups as they are encountered.

Frequently, "ideal" protocols for collecting social data are modified for reasons that are strategic (e.g., the desire simultaneously to collect data for another goal, such as population analysis) or tactical (such as weather).

3.7.4: Choosing Subjects. Although the random or systematic selection of experimental subjects is a cornerstone of statistical methodology (e.g., Sokal & Rohlf 1994, p. 393), in social analysis it is not so crucial. Clearly, if there are sets of individuals with very different social behavior, we need data on all of them, but it does not matter much if we gather relatively more data on some than on others. Hypothesis tests are not very frequent in social analysis, and those that are performed are usually framed in terms of the behavior of the individuals that have been sampled (e.g., "within the sampled population of individuals, males form larger groups than females"), in which we assume that we have obtained a random sample of the behavior of the sampled individuals, not that the sampled individuals are a random selection from the entire population.

Thus, when choosing the cave in which to identify roosting bats or which member of a captive population to begin a focal individual follow, we could use a random numbers table, but we could also use other criteria. Caves or focal individuals or classes of animals (Section 3.6) could be chosen in rotation, or a special focus could be placed on those that are deemed particularly interesting (perhaps mothers or caves with high bat densities). In more difficult research settings, subjects of surveys and follows are often chosen haphazardly, such as "the first group we come across." This is usually acceptable, even if it means that individuals with a home range near the research base are sampled more often than those who live at a greater distance.

Subject choice, however, could affect some social measures. For instance, if groups are surveyed as they are observed and large groups are more prominent than small ones, then an estimate of mean group size calculated from the data will be biased upward.

Another decision faces those who follow groups: When the group splits, which part should be followed? A rule such as "follow the largest of the daughter groups" will have no important effect on measures of dyadic relationship but would bias group size estimates. It is possible randomly to choose the daughter group that will be followed or, even better but impracticable in some circumstances, randomly pick a key individual when the group is first encountered and then, when a split occurs, follow whichever daughter group contains the individual.

3.7.5: Sampling Protocols. Altmann (1974) and others (e.g., Martin & Bateson 2007, pp. 48–61; Lehner 1998, pp. 195–210; Mann 1999) list a number of sampling protocols, such as "ad libitum," "focal-animal," "all-event," "predominant activity," "point," "scan," "1:0," or "sequence." These have different advantages, disadvantages, and recommended uses. In most formulations, the relative merits of the sampling protocols are confounded with follow protocols (discussed earlier). Here and in Table 3.4, I adapt the standard terminology and recommendations for the collection of social data, indexing choices by the follow protocol (surveys, individual follows, or group follows) and type of social measure being collected (interactions, associations or groups):

Surveys, recording associations or groups. Here the sampling is an instantaneous scan and usually 1:0 (a dyad is or is not associated, or are members or not members of the same group). Sometimes, however, individual or dyadic behavioral state data are collected. These could be ordinal or continuous data or categorical data with several states. They can be used to produce associations in subsequent analysis. For instance, locations can give nearest-neighbor data, whereas behavioral state and movement measures allow synchronicity to be assessed. In the simplest and most common format, the members of a group are noted during each survey of each group.

Individual follows, recording interactions. Ideally, each interaction involving the focal animal is recorded together with the time, type, and identity of the interactant. This is sometimes called "all-event" sampling. Simplifications include omitting the time information but recording the order of interactions (this becomes "sequence sampling" under some definitions), simply counting all interactions between each dyad ("sociometric matrix"), or recording whether there was an interaction between the focal animal and each other individual ("1:0 sampling"). I recommend that time, interaction type, and interactant identity be recorded if possible. Some measure of effort is required for most analyses, usually the time or number of sampling periods (Section 3.9) spent observing each individual.

Individual follows, recording associations or groups. There are several possible sampling protocols for association measures during individual follows. The associates of the focal individual can be recorded at regularly spaced instants (e.g., "who was the focal individual associated with at 12:05?," also called "point sampling"), during regular intervals (e.g., "who was the focal individual associated with between 12:05 and 12:10?," a form of "1:0 sampling"), or when they changed (e.g., "A became associated with the focal individual at 12:07," a form of "all-event sampling"). As long as the intervals used are not greatly longer than the rates of disassociation, it probably makes little difference to the results which of these is used. As with the survey protocols, individual data (e.g., identities and positions relative to the focal animal) can be recorded for all nearby individuals using any of these methods and then used later to produce one or more association measures with the focal animal (such as nearest neighbor).

Group follows, recording interactions. This will only be possible in rare cases in which the group is small and easily viewed and interaction rates are low, but in such cases, "event sampling" (in which events are the interaction types) is appropriate and efficient.

Group follows, recording associations or groups. The possibilities are similar to those available for association measures during individual follows listed previously, but with a few additional options. The simplest, and probably most frequently used, sampling protocol is simply to list group membership at regularly spaced sampling points ("point sampling") or whenever it changes ("event sampling"). Alternatives are to note associations within the group, if association is defined other than by membership of the followed group, either directly (e.g., subgroup membership) or by recording individual data such as position and behavioral state that can be used to derive associations later. Sometimes, these data will only be collected for a subset of the group.

It is often possible and desirable to combine sampling protocols. For instance, interactions and associations can both be recorded during individual follows. Sometimes, the data collected can be used to derive two or more association measures, such as "behavioral coordination" and "within x body lengths." Finally, the maligned ad libitum sampling method (basically field notes) should be used to record unusual but important behavior, such as fighting or mating, whether or not the focal individual is a participant and whether or not the behavior occurs during a survey or follow (Altmann 1974).

3.7.6: Effects of Observers. It is important both ethically and scientifically to minimize the effects of observation. Disturbed animals may form larger or smaller groups or increase or decrease their rates of association or disassociation (e.g., Foster & Rahs 1983; Kinnaird & O'Brien 1996), often showing antipredator-type behavior (Frid & Dill 2002). Martin and Bateson (2007, pp. 17–18) and Lehner (1998, p. 210) discuss causes and remedies for observer effects. Similarly, effects on behavior caused by individual identification (Section 3.5) and the collection of data for classifying animals (Section 3.6) should be minimized.

3.7.7: Nonobservational Data. It has been tacitly assumed throughout this section that interaction or association data are collected through visual observation, which may be real time or by analysis of video or still images. There are, however, other sensory modes. Interactions can sometimes be heard. Associations can be measured in a large range of ways (Table 3.2), including the co-occurrence of natural (e.g., DNA analysis of discarded body tissue or feces) or artificial (e.g., PIT tag) individual markers, as well as through the products of nonsocial analyses (such as the overlap of ranges).

3.8 Data Formats

Database and spreadsheet software are almost essential for storing social data (Section 2.9), but what format should be used? In this section, I recommend formats that either allow relatively simple manipulation in spreadsheet programs such as Excel or are suitable for my software package, SOCPROG. Other specialized software packages, such as UCINET and MatMan (Section 2.9), assume some processing of the raw data into similarity or dissimilarity matrices (Section 2.5). The preferred format may depend on whether interactions, associations, or groups are recorded directly or are derived from other recorded measures. Finally, I suggest a format (the SOCPROG format) for entering supplemental data, such as age or sex, that directly or indirectly can be used to allocate individuals to classes (Section 3.6).

First, a few preliminaries. I suggest that dates and times be combined in one field using the database or spreadsheet date–time format. Second, changes and ambiguity in field (column) formats can cause problems in

Table 3.5 Example of Interaction Data with Two Asymmetric (X and Y) and One Symmetric (M) Interaction Types Coded in Linear Mode for Situations in Which All Individuals Are Observable

Date and time	Type of interaction	Actor/recipient	Interaction no.	ID
12/9/89 9:01	X	0	1	A1
12/9/89 9:01	X	1	1	A9
12/9/89 9:22	X	0	2	A14
12/9/89 9:22	X	1	2	A15
12/9/89 12:10	Y	0	3	B8
12/9/89 12:10	Y	1	3	A11
12/9/89 12:17	X	0	4	A13
12/9/89 12:17	X	1	4	A20
12/9/89 15:32	M	1	5	A4
12/9/89 15:32	M	1	5	A7
12/9/89 15:44	X	0	6	B12
12/9/89 15:44	X	1	6	A17
12/10/89 9:09	Y	0	11	A19
12/10/89 9:09	Y	1	11	A1
12/10/89 9:40	M	1	12	A9
12/10/89 9:40	M	1	12	A14

Asymmetric association data can be coded similarly.

analyses within Excel (and probably other spreadsheet software) as well as when the data are exported into other programs (such as SOCPROG). Therefore, I suggest that one not identify some individuals (or behavior types or classes) by numbers (such as "1453") and others alphanumerically (such as "53c"); one should just use numbers or alphanumeric codes throughout each field, whichever is more appropriate.

Data are usually stored so that rows represent observations, and columns (fields) the circumstances of the observation, what was observed, and who was observed. In SOCPROG, the final field gives the identities of the observed individuals, and I stay with this convention in the examples given later. I distinguish three ways of coding social data:

1. *Linear mode* (e.g., Table 3.5), in which each row corresponds to one observation of one individual. This is a SOCPROG format.

2. *Dyadic mode* (e.g., Table 3.6), in which each row corresponds to an observation of an association or interaction of a dyad. Thus, there are two identity fields representing the two identities in the dyad. This is a particularly useful format for asymmetric interactions or associations. Occasionally, the two identities may be of the same individual, as when an individual grooms itself (e.g., Table 3.6) or the presence of

Table 3.6 Example of Asymmetric Interaction or Association Data Coded in Dyadic Mode for Situations in Which All Individuals Are Observable[a]

Date and time	Groomer ID	Groomee ID
1/1/00 9:49	131	202
1/1/00 14:54	142	155
1/1/00 15:41	176	194
1/2/00 9:11	194	202
1/2/00 9:41[a]	100	100
1/2/00 10:09	6	162
1/3/00 10:35	100	188
1/3/00 11:03	196	202
1/3/00 14:32	6	162
1/3/00 17:40	155	89
1/4/00 7:16	196	202
1/4/00 13:17	131	3
1/4/00 16:15	155	89
1/5/00 6:00	51	89
1/5/00 15:57	162	100
1/5/00 17:55	131	3
1/11/00 7:19	188	127
1/11/00 10:09	89	45
1/12/00 7:14	89	45
1/12/00 9:01	162	100

[a] Including one case in which the interaction is of an animal with itself.

a noninteracting or nonassociating individual needs to be noted (see later discussion). This is a useful format for processing in Excel or other spreadsheet programs, for instance, by using "pivot tables" to produce counts of interactions or associations.

3. *Group mode* (e.g., Table 3.7) in which observations of one, two or more than two individuals are represented on each row, and one field gives all the identities of the individuals observed in the group. This is a SOCPROG format, and is compact, using less computer space and memory than individual or dyadic mode to store the same data.

Dyadic and group mode data can always be converted to linear mode data,[2] and linear mode data can usually be converted to dyadic mode. Linear and dyadic mode data cannot necessarily be converted into group mode, however, because linear and dyadic mode data are not necessarily symmetric and transitive, a requirement for group mode data (if A and

2. SOCPROG can convert group mode data into linear mode.

Table 3.7 Example of Coding Symmetric Association Data with
One Association Type Such as Group Membership, Coded in Group
Mode, Collected from Surveys

Date and time	Associating IDs
1/1/00 9:49	6 13 20
1/1/00 14:54	15
1/1/00 15:41	17 19
1/2/00 9:11	20
1/2/00 9:11	10 18
1/2/00 10:09	6 16
1/3/00 10:35	5 10 18
1/3/00 11:03	20
1/3/00 14:32	6 10 16 20

B interact/associate and B and C interact/associate, this does not necessarily imply that A and C interact/associate).

3.8.1: Coding Interaction Data. Coding interaction data is not always
straightforward. The simplest case occurs when the whole population
and all their interactions of certain types are observed and all interaction types are symmetric, so that there is no ordering to the interaction.
Then, we can use dyadic or group mode data storage, with each row
representing an interaction. The fields will usually contain date/time,
type of interaction if more than one is observed and the identities of
the interactants, as in Tables 3.6 and 3.8. Additional fields may contain
information such as place and intensity of the interaction. Interactions
involving three or more individuals can also be coded in this way in
group mode simply by having more than two individuals recorded in
the ID field, or in dyadic mode by having, for three individuals, three
rows representing all three dyadic interactions.

If interactions are not symmetric, as in grooming (A may groom B
without B grooming A) or fight outcomes, then we can use group mode
data storage, but the first individual listed is considered the actor and
the second (or perhaps all of the others if more than two individuals
are listed) the receivers. Dyadic mode is particularly well suited for this
situation (e.g., Table 3.6), or it may be best to use linear mode to code
the data. One field represents the interaction type and another whether
an individual is the actor or recipient, and a third distinguishes the different interactions, as in Table 3.5, which codes a mixture of symmetric
and asymmetric interaction types. For each observed asymmetric interaction one individual has a "o" in the actor/recipient field and another
individual a "1." For symmetric interactions, both individuals have a

Table 3.8 Example of Interaction Data with Five Symmetric Interaction Types
Coded in Group Mode for Situations in Which All Individuals Are Observable

Date and time	Interaction type	Interactant IDs
1/1/00 9:49	A	13 20
1/1/00 14:54	A	14 15
1/1/00 15:41	A	17 19
1/2/00 9:11	F	19 20
1/2/00 9:41	F	10 18
1/2/00 10:09	A	6 16
1/3/00 10:35	A	10 18
1/3/00 11:03	D	19 20
1/3/00 14:32	A	6 16
1/3/00 17:40	A	15 8
1/4/00 7:16	D	19 20
1/4/00 13:17	A	13 3
1/4/00 16:15	A	15 8
1/5/00 6:00	A	5 8
1/5/00 15:57	B	16 10
1/5/00 17:55	F	13 3
1/11/00 7:19	C	18 12
1/11/00 10:09	A	8 4
1/12/00 7:14	C	8 4
1/12/00 9:01	B	16 10

"1." This format can also be extended for triadic interactions or those including more than three individuals.

An additional, but very important, consideration in most circumstances is the coding of control data. In addition to recording the observed interactions, we need to know for which individuals we could have recorded interactions had they taken place, so that, for instance, rates of interaction per unit time can be calculated for each dyad. Thus, in cases such as focal animal or group follows, where not all members of the population are being observed all the time, effort data must be coded in some way. This can be done by including, at least once per sampling period, data on "null interactions" that simply note the animals that could have interacted. For focal group follows, this can be achieved in group mode format. With individual follows, however, linear or dyadic mode will usually be needed as if A is being followed and B and C are also being observed such that they could have interacted with A; then we need to record the possibility of AB and AC interactions, but not BC ones. Tables 3.9 to 3.11 give, respectively, examples of group, dyadic, and linear mode interaction records with control data.

3.8.2: Coding Direct Association Data. When symmetric associations are recorded directly, the data can be coded using dyadic mode, as in Table

Table 3.9 Example of Interaction Data with Two Symmetric Interaction Types Coded in Group Mode, Including "Null" Interaction Effort Data ("0"), as Might Be Obtained from a Focal Group Follow

Date and time	Interaction type	Interactant IDs
1/1/00 9:40	0	13 18 20
1/1/00 9:45	0	13 18 20
1/1/00 9:49	A	13 20
1/1/00 9:50	0	13 14 20
1/1/00 9:55	0	13 14
1/1/00 14:45	0	14 15 18 20
1/1/00 14:50	0	14 15 18 20
1/1/00 14:54	A	14 15
1/1/00 14:55	0	15 18 20
1/1/00 14:58	B	15 20
1/1/00 15:00	0	15 18 20

Table 3.10 Example of Asymmetric Interaction Data Coded in Dyadic Mode, Including "Null" Data—Individuals Who Could Have Groomed but Did Not—as Might Be Obtained from a Focal Follow of #131

Time	Grooming?	Groomer ID	Groomee ID
06:22:00	Yes	131	202
06:27:00	Yes	131	155
06:27:00	No	176	131
06:32:00	No	176	131
06:37:00	Yes	131	176
06:37:00	No	131	162
06:42:00	No	131	188
06:47:00	No	131	202
06:52:00	Yes	6	131
06:52:00	Yes	131	89
06:57:00	No	131	202
06:57:00	No	131	89

3.6, or group mode, as in Table 3.7, in which each row corresponds to animals that are associated with each other. If more than two identifications are noted in group mode, then each is assumed to have been associated with all the others. The coding is basically the same whether the data come from surveys, group follows, or individual follows and whatever sampling protocol is used. It is important, however, that all individuals observed within a sampling period are noted, with an individual that was not associated with any other being indicated by a single identification in a row in group mode (as in Table 3.7), or, in dyadic mode, as an association of an animal with itself. With asymmetric associations (such as nearest-neighbor measures), then linear or dyadic mode coding is required, as in Tables 3.5, 3.6, 3.10, and 3.11.

Table 3.11 Example of Interaction Data with Two Asymmetric (X and Y) Interaction Types plus "Null" Data ("0") Coded in Linear Mode as Might Result from Focal Individual Follows

Date and time	Type of interaction	Actor/Recipient	Interaction#	ID
12/9/89 8:55	0	1	1	A1
12/9/89 8:55	0	1	1	A9
12/9/89 8:55	0	1	2	A1
12/9/89 8:55	0	1	2	B10
12/9/89 9:00	0	1	3	A1
12/9/89 9:00	0	1	3	A9
12/9/89 9:00	0	1	4	A1
12/9/89 9:00	0	1	4	B10
12/9/89 9:01	X	0	5	A1
12/9/89 9:01	X	1	5	A9
12/9/89 9:05	0	1	6	A1
12/9/89 9:05	0	1	6	A9
12/9/89 9:20	0	1	7	A14
12/9/89 9:20	0	1	7	A15
12/9/89 9:25	0	1	8	A14
12/9/89 9:25	0	1	8	A15
12/9/89 9:30	Y	0	9	A14
12/9/89 9:30	Y	1	9	A15
12/9/89 9:35	0	1	10	A14
12/9/89 9:35	0	1	10	A15
12/9/89 9:40	0	1	11	A14
12/9/89 9:45	0	1	12	A14

A1 and A14 are the focal individuals.

3.8.3: Coding Indirect Association Data. Sometimes, associations are not recorded directly but are inferred later. In this case, data are generally recorded in linear mode. Fields may include date/time, position (one, two, or possibly three dimensional), heading, or behavior (events or states). Then one can derive association measures such as "nearest neighbor," "dived within 30 seconds of one another," or "within three body lengths and heading the same direction ($\pm 30°$)." SOCPROG can usually produce such association measures reasonably easily. Table 3.12 shows an example of such data.

3.8.4: Coding Group Data. Group data are the simplest to code, usually in group mode, as in Table 3.7, in which each row corresponds to a group. Once again, it is important that single animals are entered, as a row containing just one ID.

3.8.5: Coding Supplemental Data. For social analysis, in addition to data on interactions or associations, we generally use attributes of individuals to place them into classes (Section 3.6). Individual attributes can be used to calculate nonsocial relationship measures, such as age differences

Table 3.12 Example of Data Coded so That Associations Can Be Derived Later

Date and time	Position on branch	Branch no.	ID
12/9/89 9:01	12	1	A1
12/9/89 9:01	27	1	A9
12/9/89 9:01	31	1	A14
12/9/89 9:01	37	1	A15
12/9/89 12:10	5	2	B8
12/9/89 12:10	15	2	A11
12/9/89 12:17	6	3	A13
12/9/89 12:17	9	3	A20
12/9/89 15:17	21	3	A4
12/9/89 15:17	25	3	A7
12/9/89 15:17	29	3	B12
12/9/89 16:40	31	4	A17
12/10/89 9:09	19	5	A19
12/10/89 9:09	25	5	A1
12/10/89 9:09	31	5	A9
12/10/89 9:09	50	5	A14

The identities of birds perching on surveyed branches are recorded together with the position on the branch (in centimeters from the trunk of the tree). Associations such as "nearest neighbor" and "on same branch and within 15 cm" can be calculated from these data.

Table 3.13 Example of Supplemental Data That Assign Individuals to Classes (e.g., Sex), Can Be Used to Derive Classes (e.g., from Age, One Can Derive Age Classes), or Can Be Used to Produce Nonsocial Relationship Measures (Such as Haplotype Similarity)

ID	Sex	Age (yr)	Haplotype
1	M	15.5	A
2	M	2.7	H
3	F	5.8	H
4	M	14.5	G
5	M	20.8	F
6	F	9.7	A
7	F	7.4	F
8	F	24.6	G
9	M	6.1	H
10	F	17.2	A
11	M	11.7	A
12	M	17.7	F
13	F	11.7	A
14	M	4	B
15	M	15.7	C
16	F	0.3	A

or genetic relatedness (Section 4.2). Class allocations, or data used to produce them, can also be stored in spreadsheet or database format. Table 3.13 illustrates the format used by SOCPROG and UCINET. The first column (field) is a list of identification names or numbers, and the subsequent columns (fields) give information such as sex, age, or haplotype for the corresponding individual. If a spreadsheet program such as Excel is being used, supplemental data can be stored in the same file as the social data but using separate worksheets. When we have both social data and supplemental data, it is perhaps even more useful to use a relational database such as Access with linked relationships, which makes it simple to both change and view aspects of the data.

3.9 Sampling Periods

Time has two important roles in social analyses. First, because the temporal patterning of interactions and associations is one of the key attributes of a relationship (Hinde 1976), temporal methods should play a key role in the analysis (Sections 4.6 and 5.5). At a more basic level, for almost all statistical techniques, we need to define a "sampling period"—the temporal units of the analysis. Thus, in each sampling period, we may produce counts of dyadic interactions or abstract whether a dyad was associated or not.

There are a number of considerations in selecting a suitable sampling period, including natural breaks in the sampling scheme, the rate of data collection, and independence of neighboring periods. For instance, if sampling is only carried out in darkness or only in daylight, then a sampling period of a day has a natural break and may be appropriate. A sampling period so short that there are few data collected within it (e.g., few interactions observed) is rarely useful. At the other extreme, valuable information is lost if the sampling period is so long that, for instance, almost all individuals in the population have associated with each other during each period. If association data are being collected and associations in consecutive intervals are almost always identical, then the sampling period is probably too short, whereas if they are almost uncorrelated, then the sampling period may be too long. For most analyses, statistical independence between neighboring sampling periods is neither needed nor desirable because we are interested in how dyadic relationships change over a range of time scales (Section 4.6). However, there are exceptions. Some permutation tests (Bejder et al. 1998) and estimates of the power of social analyses (Section 3.11), assume independent sampling periods, independent in the sense that the data from neighboring sampling periods are no more alike than those from well-separated periods. Thus, for different analytical techniques, it may be appropriate to divide the data into sampling periods of different durations. Occasionally, it may be useful to use sampling periods defined by a measure other than time, such as the field study or survey.

3.10 *Attributes of Data Sets*

When individuals have been identified and data on their interactions or associations recorded, we have a data set from which, potentially, we can analyze social structure. Such data sets vary widely over a large range of attributes, as indicated by the example studies listed in Table 1.3. Some of the most basic features of data sets and study populations are the number of identified individuals; the proportion of the population that is identifiable (possessing marks, tags, or other attributes used for individual identification); the proportion of the population that has been identified in the study; whether individuals are classifiable by age, sex, or other attributes; the length of the sampling period; the number of sampling periods; their temporal pattern; the proportion of the population identified during each sampling period; the number of study areas; their geographical relationship; the migration rates between study areas; movement and spatial structure within study areas; the number and type (presence/absence, ordinal, etc.) of interaction and association measures; and whether there are missing or incomplete data (e.g., some measures not always collected).

These attributes allow data sets to be allocated to general types. Here are some axes by which data sets might be classified and suggested classes:

> *Size of study population*: "Small," less than 20 identified individuals; "intermediate," 21 to 100 identified individuals; "large," more than 100 identified individuals.
>
> *Rate of identification*: "Sparse," less than 10% of study population identified during each sampling period; "intermediate," 10% to 80% of study population identified during each sampling period; "complete," greater than 80% of study population identified during each sampling period.
>
> *Number of sampling periods during which a dyad is observed associated*: "Few," less than 1 mean observed associations per dyad; "some," 1 to 10 mean associations per dyad; "many," more than 10 mean associations per dyad.
>
> *Associations per individual*: "Few," less than 10 mean observed associations per individual over all other members of the population; "some," 10 to 100 mean associations per individual; "many," more than 100 mean associations per individual.

Population closure: "Closed," no birth, death, or emigration during the study; "open," some individuals enter or leave the population during the study.

Information on individuals: "Undifferentiated," no class information; "gender," sex, but no other class data, available for each individual; "detailed," gender plus other individual data, such as age, available for each individual.

Behavioral measures recorded: "Univariate," just one interaction or association measure available; "bivariate," two interaction or association measures available; "multivariate," more than two measures available.

Length of data set: "Short," less than 20 sampling periods; "medium," 20 to 100 periods; "long," more than 100 periods.

3.11 How Large a Data Set Is Needed for Social Analysis?

This book is principally concerned with deriving a model of social structure from observations of the social behavior of animals. As far as possible, the output model of social structure should correspond to its real nature. The closer this match, the better. Using the best methods improves the input–output match, and Chapters 4 and 5 are principally about the utility of different methods. Even if the ideal analytical techniques are employed, however, the output model will likely have little basis in reality if insufficient data are input. So how large a data set is needed for social analysis? In this section, I look at the precision and power of social analysis (Whitehead In press-a). Three subsections are concerned with the precision of dyadic relationship measures, the accuracy of representations of social structures, and the power of tests of null hypotheses, respectively.

3.11.1: Precision of Relationship Measures. In Chapter 4, I describe several relationship measures, primarily interaction rates (Section 4.4) and association indices (Section 4.5). These indicate the strength and nature of a relationship between a pair of individuals.

An interaction rate measures how frequently interactions occur between a pair of individuals. If interactions can be considered independent then, from Equations (4) and (5) in Chapter 4, the coefficient of variation (CV; standard error divided by mean) of an estimated interaction rate between individuals A and B is approximately $1/\sqrt{n_{AB}}$, where n_{AB} is the number of observed interactions between them.

Table 3.14 Precision of Relationship Measures as Indicated by the Expected Coefficient of Variation (CV) of Interaction Rates and Association Indices as a Function of the Number of Observed Interactions or Associations and (for Association Indices) the Association Index

Number of observed interactions or associations	CV of interaction rate	CV of association index if association index is:				
		0.1	0.3	0.5	0.7	0.9
5	0.45	0.42	0.37	0.32	0.24	0.14
10	0.32	0.30	0.26	0.22	0.17	0.10
20	0.22	0.21	0.19	0.16	0.12	0.07
40	0.16	0.15	0.13	0.11	0.09	0.05
80	0.11	0.11	0.09	0.08	0.06	0.04

An association index, α_{AB}, estimates the proportion of time that a pair of individuals A and B is in association. From Equation (6) in Chapter 4, its CV is approximately $\sqrt{((1 - \alpha_{AB})/x_{AB})}$, where x_{AB} is the number of observed associations (actually number of sampling periods with an observed association) between A and B.

Using these formulas, Table 3.14 presents estimated CVs of interaction rates and association indices. Unless the association index is close to 1, one needs at least 15 independent observations of interactions or associations to lower the CV of the relationship measure below 0.15, thus giving 95% confidence intervals in the relationship measures of $\pm\sim30\%$ of the mean (because 95% confidence intervals are roughly twice the SE, and CV = SE/mean). To obtain even more precise interaction rates or association indices, many independent observations are required.

3.11.2: Accuracy of Social Representations. In Chapter 5, I show how estimated relationship measures among members of a population can be used to construct representations and models of social structures. Given that each estimated relationship measure will likely have an error, however, how accurate is a representation built on many such imperfect measures? This has not been fully explored, but in the case of association indices assembled into matrices of association indices (such as those in Tables 2.5, 4.16, and 4.17), I have made a start (Whitehead In press-a).

The measure of accuracy used is the correlation coefficient between the true association indices—what proportion of time a pair are actually associated—and their estimated values, the association indices. A high correlation, with r near 1.0, indicates an excellent representation; $r \sim$ 0.8 indicates a good representation; and $r \sim 0.4$ indicates a somewhat representative pattern. I have shown (Whitehead In press-a) that this correlation can be estimated, in a fairly unbiased fashion, from

$$r = \frac{S}{CV(\alpha_{AB})} \tag{1}$$

where S is the social differentiation, the estimated CV of the true association indices (Section 5.1), and $CV(\alpha_{AB})$ is the CV of the estimated association indices. Briefly, S indicates the variability of association indices within a population: if S is close to 0, the relationships within the population are homogeneous; if S is near or greater than 1, they are very varied. S can itself be estimated in two ways described by Whitehead (In press-a) and in Appendix 9.4.

Equation (1) allows post hoc estimation of the accuracy of social representations based on calculated association indices. For instance, for the 63 northern bottlenose whales (*Hyperoodon ampullatus*) whose data are summarized in Tables 4.19 and 5.1, the estimated correlation coefficient between the true and estimated association indices is 0.22, which is not good and suggests that, in this case, representations of the matrix of association indices will not reflect reality to any great extent.

We do not know $CV(\alpha_{AB})$ until the data are collected, so Equation (1) does not help directly with addressing the question of how much data are needed to achieve an "accurate" social representation. By making the simplifying assumptions that the observations of associations are Poisson distributed and that effort is equally concentrated on all dyads, we can come up with a formula that is useful for predicting the accuracy of social representations (Whitehead In press-a):

$$r = \sqrt{\frac{1}{1 + \frac{1}{S^2 \cdot G}}} \tag{2}$$

where G is the mean number of associations observed per dyad This relationship allows prediction of the correlation between the true and estimated association indices, as is done in Table 3.15. Equation (2) assumes that effort is equally concentrated on all dyads. If this is not the case, so that more effort is devoted to some dyads than to others, then the power to assess the true association index will be reduced.

The amount of data needed to give a "somewhat representative" pattern of relationships within a population, indicated by $r = 0.4$, or a "good" representation, indicated by $r = 0.8$, varies greatly with the

Table 3.15 For Different Levels of Social Differentiation, Estimates of the Quantity of Data Required, as Expressed by the Mean Number of Observed Associations per Dyad, to Obtain a "Somewhat Representative" Pattern of Social Relationships within a Population or a "Good" Representation from Equation (2) and to Reject the Null Hypothesis of No Preferred or Avoided Companions Using the Bejder et al. (1998) Test from the Relationship $S^2 \cdot g' > 5$

Social differentiation (S)	Mean number of observed associations per dyad (G):		Mean observed associations per individual (g') for probable rejection of null hypothesis of no preferred/avoided companionship
	For "somewhat representative" picture of social structure; $r = 0.4$	For "good" representation of social structure; $r = 0.8$	
0.05	76.19	711.11	2000
0.2	4.76	44.44	125
0.8	0.3	2.78	7.8
2.5	0.03	0.28	0.8
10	0.002	0.02	0.05

social differentiation (Table 3.15). With a poorly differentiated population ($S < \sim0.2$), many associations are needed per dyad to achieve even a "somewhat representative" pattern, whereas when social differentiation is high, the data requirements are much less. About 10 times as much data are required for a "good" representation, $r = 0.8$, as a "somewhat representative" one with $r = 0.4$ (Table 3.15).

3.11.3: Power of Tests of Null Hypotheses of No Preferred Companionship. Social analysis can also be approached from a hypothesis-testing perspective. A frequently useful null hypothesis is that individuals have no preferences for social partners, with the alternative that there are preferred and/or avoided associations between some pairs of individuals. In Section 4.9, I describe a permutation test introduced by Bejder et al. (1998) that tests for preferred or avoided association among individuals when dyadic association is defined using group membership (Section 3.4), as well as several extensions of this test. These are important techniques in social analysis. However, the power of such tests is unknown: How much data are needed to detect a particular degree of social preference/avoidance?

In simulated data sets, the null hypothesis of no preferred companions was usually rejected (at $P < 0.05$) if $S \cdot g' > 5$ and not rejected ($P < 0.05$) if $S \cdot g' < 5$, where g' is the mean number of observed associations per individual, not per dyad as in the case of G (Whitehead In press-a). Thus, to detect preferred companionship in a data set with low social differentiation ($S = \sim0.05$), we need a mean of about 2,000 observed associations per individual; with medium social differentiation ($S = \sim0.2$), we need about 125; with high social differentiation ($S = \sim0.8$), we need about 8; and with extreme social differentiation ($S = \sim10$), we need just 0.05 (Table 3.15).

As an example, for the 63 northern bottlenose whales whose data is summarized in Tables 4.19 and 5.1, $S \cdot g' = 11.3$, which is greater than 5, suggesting the ability to detect nonuniform associations and that the null hypothesis of social homogeneity (no social differentiation) will be rejected. It is rejected ($P < 0.001$, testing by permuting the matrices of association indices, using the coefficient of association indices as a test statistic, as described in Section 4.9). This indicates that although a quite sparse data set on a fairly large number of animals (63 in this case) may be unable to provide a good representation of social structure, it can provide useful social information.

BOX 3.1 *Precision and Power of Social Analysis*

The considerations of Section 3.11 show that in most situations, quite a great deal of data are needed to give even reasonably useful portrayals of social systems. This is especially true when social differentiation is low. Only in cases of highly socially differentiated populations for which relationships among dyads vary substantially can we get by with sparse data. This indicates that many published social analyses contain representations of social systems and conclusions about social features that may have little validity because of poor analytical precision or power.

---

4 Describing Relationships

4.1 Relationships

Whereas interactions are the foundations of social structure, the relationship is its heart. A relationship may seem simple. Perhaps the relationship between two males has always between characterized by mutual antagonism. Even within this consistent hostility, however, the level of dominance may have shifted over time. Relationships may be much more complex, especially in cooperative species, with bonds developing and deteriorating over time, being tested by agonism expressed in many behavioral contexts, often asymmetrically. de Waal (1998, pp. 83–135) describes in dramatic detail the changes in a complex relationship between two adult male chimpanzees (*Pan troglodytes*) in the Arnhem Zoo. As Yeroen and Luit struggled for power, their dynamic relationships with other members of the colony, males and females, became crucial.

Relationships between individuals form and develop through experience, learning, feedbacks, and institutionalization (Hinde 1976). In social analysis, we need to describe relationships in ways that are both meaningful and tractable. This is challenging in many ways, but the challenges are circumscribed by the data.

A primary challenge is time scale. From their very definition, relationships are integrations of interaction data over time, and yet relationships change with time (Hinde 1976). When we make quantitative representations of relationships, we need to incorporate data over sufficient time periods to make them good representations but not over time periods so long that the relationship is likely to have changed substantially. Choosing suitable time periods over which to average needs a good feel for the time scales that are significant to the animals. There are also analytical techniques, such as lagged association rates (Section 5.5), that can provide guidance with this.

Once data have been collected and a sampling period chosen, the next step in a social analysis is to calculate measures that describe the interactions and associations in each sampling period, what I have called "interaction measures" (Whitehead 1997). These are often counts of interactions (Section 3.2) or records of associations (Section 3.3). The interaction measures can be abstracted over sampling periods to produce *relationship measures*, such as the mean interaction rate or association index of a dyad, that describe the content, quality, and temporal patterning of relationships (Whitehead & Dufault 1999). The heart of this chapter is a discussion of interaction and relationship measures, but before that, I consider nonsocial dyadic measures (Section 4.2) and individual attributes (Section 4.3), both of which give perspective on dyadic measures of sociality.

Table 4.1 gives an overview of much of the material in this chapter. It lists general types of relationship measures, the level or levels at which they operate (individuals, dyads, classes, or community), and where there are tests against null hypotheses.

A fundamental attribute of the methods that I advocate in this chapter and the chapters that follow it is that they adopt an animal-centered approach (Jarman 1982; Whiten 2000). They try to view a social structure from the perspective, and through the relationships, of its members rather than just as perceived by external observer.

4.2 Nonsocial Measures of Relationship

The dynamics of the social relationship between the male chimpanzees (*Pan troglodytes*) Yeroen and Luit at the Arnhem Zoo were heavily influenced by the relative ages and consequent stamina of the participants. Yeroen was older, but he tired more quickly (de Waal 1998, pp. 50–53). Most fundamentally, both were mature males.

There is a range of possible and useful dyadic nonsocial measures of relationship. Nearly all of the class attributes of individuals (Section 3.6) can be used to produce such measures. Some are 1:0 or same/different

Table 4.1 Summary of Relationship and Related Measures

	Individuals	Dyads	Evaluated for: Class	Between classes	Whole community
Interaction	Mean interaction rate (section 4.3*)	Mean interaction rate (sections 4.4, 4.8*)	Mean interaction rate	Mean interaction rate between classes (section 4.11)	Mean interaction rate (section 5.1)
Association	Gregariousness (section 4.3*)	Association index (sections 4.5, 4.9*)	Gregariousness of class (sections 4.3, 4.11)	Gregariousness between classes (sections 4.3, 4.11); mean association index between classes (sections 4.9,* 4.11)	Mean gregariousness (typical group size, section 5.1); mean association index
Temporal change in interaction or association rates	—	Development or lagged rates (section 4.6)	—	Between-class lagged rates (section 4.6, 4.11)	Lagged rates (section 5.5)
Dominance	Dominance rank (section 4.3)	Dominance (section 4.8)	Class dominance hierarchy	Class dominance (section 4.11)	Dominance hierarchy (section 5.4*)
Symmetry	—	Symmetry of interaction rates (section 4.8*)	—	Mean symmetry of interaction rates between classes (section 4.11)	Overall symmetry of interaction rates; mean directional consistency index (section 5.1)
Reciprocity	Relative reciprocity in asymmetric interactions involving one individual (section 4.8*)	—	—	Relative reciprocity in asymmetric interactions between classes (section 4.8)	Overall relative reciprocity in asymmetric interactions (section 4.8*)
Bonds	—	"Significant" values on two or more independent relationship measures (section 4.10)	—	—	—

An asterisk indicates that a significance test against a null hypothesis of no individual or class effect is discussed in the text.

dyadic measures, including same/different sex, age class, reproductive state, subspecies, morph, or matriline (i.e., with a recent common ancestor in the maternal line). Alternatively, we can have several categorical dyadic states for each combination of individual classes, such as, for sex classes, F-F, M-M, and M-F. Nonsocial dyadic measures can also be quantitative, such as the difference in age between the two individuals.

Among the most commonly used nonsocial measures of relationship are those that indicate kinship or genetic relatedness. As discussed in Section 7.3, kinship is believed to be one of the drivers of sociality, and a general prediction of kinship theory is that the social relationship in a dyad should be correlated with its members' genetic relatedness. If the genealogy is known, kinship (an estimate of the probability that individuals have a gene in common through common recent ancestry) between any pair of individuals can be calculated. Alternatively, kinship is indicated by genetic relatedness estimated using molecular genetic techniques, frequently microsatellites (van de Casteele et al. 2001) (e.g., Table 4.2). Dyadic kinship measures usually estimate the proportion of genes shared through common recent ancestry, and so they should range from 0 to 1, although, in practice, there are frequently small negative relatedness estimates. A molecular genetic 1:0 measure of matrilineal relatedness that is methodologically simpler to derive is whether the pair of individuals does or does not share the same mitochondrial DNA haplotype. Individuals from the same matriline should have the same mitochondrial DNA haplotype; those from different matrilines may not.

Other important nonsocial dyadic measures include spatial and/or temporal range overlap (e.g., Gompper et al. 1998). The former is often derived from habitat occupancy data using geographic information systems.

4.3 Social Attributes of Individuals, Including Gregariousness

In the following sections of this chapter, I consider dyadic measures of social relationship; to place these in perspective, however, we need to think about social attributes of individuals. Just as individual, nonsocial class attributes (Section 3.6) can be used to produce nonsocial measures of relationship (Section 4.2), so social attributes of individuals affect measures of social relationship. For instance, in the power struggle between the chimpanzees (*Pan troglodytes*) Yeroen and Luit at the Arnhem Zoo, a crucial factor was the greater sociability of Luit (de Waal 1998, p. 53), which allowed him to form alliances more easily.

An important social attribute of individuals in many societies is *dominance rank*, the ranking of an in individual, within its community, in

Table 4.2 Genetic Relatedness among Male Bottlenose Dolphins (*Tursiops* spp.), Members of a "Superalliance," in Shark Bay, Australia, as Estimated Using 12 Microsatellite Loci

	LAT	GRI	VAX	KRI	MYR	WOW	HOB	WBE	HOR	AJA	PIK	ANV
GRI	-0.24											
VAX	0.02	0.08										
KRI	0.02	-0.04	-0.19									
MYR	-0.27	0.44	-0.03	-0.11								
WOW	0.22	0.11	0.32	-0.10	0.10							
HOB	-0.04	0.11	-0.17	-0.13	-0.08	-0.12						
WBE	0.15	0.07	-0.08	0.08	-0.08	0.23	0.13					
HOR	-0.08	0.21	-0.14	-0.23	0.18	0.12	0.11	0.26				
AJA	-0.24	0.23	-0.04	-0.16	-0.01	-0.16	0.07	0.25	0.32			
PIK	-0.11	0.35	-0.07	0.04	0.02	-0.05	0.09	0.60	0.21	0.27		
ANV	-0.05	-0.23	-0.39	-0.39	-0.21	-0.13	-0.41	0.11	0.11	0.02	-0.06	
VEE	0.14	0.02	0.15	-0.11	-0.08	0.00	-0.09	-0.05	0.06	0.01	-0.17	-0.17

A relatedness of 0.0 is the average among members of the wider population; 0.25 is expected for half-brothers. Although relatedness theoretically varies between 0.0 and 1.0, estimates of relatedness using molecular genetic methods often produce small negatives values.
From Krützen et al. (2003).

its ability to consistently win repeated agonistic encounters with other members of the community (Drews 1993). Yeroen and Luit competed for dominance rank. Although dominance rank is an attribute of individuals, it is actually the result of dyadic interactions and is estimated using interaction data (for methods, see Sections 4.8 and 5.4). An alternative to the standard integer dominance rank is the *dominance index,* which expresses an individual's ability to dominate others. Some dominance indices are described in Section 5.4.

Pepper et al. (1999) draw attention to *gregariousness,* which measures an individual's tendency to form associations. An individual with high gregariousness has more and/or stronger relationships than a less gregarious member of its community. Once association has been defined (Section 3.3), gregariousness can be measured simply as the mean number of associates possessed by an individual, or the sum of all dyadic association indices (Section 4.5) involving a particular individual (usually excluding the 1.0 of the individual with itself). If association is defined using group membership, then a measure of gregariousness for an individual is its typical group size (Section 3.4) minus one, the mean group size that it experiences (Underwood 1981; Pepper et al. 1999)—the average number of other individuals in the same group as an individual.

With individuals assigned to classes, we can consider gregariousness within and between classes (Underwood 1981; Pepper et al. 1999). A female possesses a gregariousness in its relationships with other females, as well as a separate gregariousness with males, estimated simply from the means of the number of other females or males in groups containing the individual. The former could be relatively low and the latter relatively high. Such a female would generally be found associating with more males but fewer females than an average female.

In an analogous manner to the way in which measures of gregariousness can be calculated from association indices, we can measure the average rate for any type of interaction for individuals or classes of individual, within or between classes. For instance, here are the percentages of total time spent grooming for six adult male chimpanzees studied at Gombe in 1978 (Goodall 1986, p. 395):

Humphrey	7.5%
Evered	34.5%
Figan	15.0%
Satan	36.0%
Jomeo	16.0%
Sherry	15.0%

These measures of social attributes of individuals are usually informative in their own right, but they also have another important role: Differences in rates of association (gregariousness) and interaction among individuals and classes of individual affect dyadic interaction and association measures (e.g., two very gregarious individuals are likely to have a high association index), and these may need to be controlled for when examining dyadic relationships. This is covered in Sections 4.8 and 4.9.

4.3.1: Tests for Differences in Gregariousness or Interaction Rates among Individuals. Permutation tests (Section 2.4) can be used to test the null hypothesis that all individuals in the community have the same gregariousness against the alternative that gregariousness varies among individuals.

If association is defined using group membership, then a test for differences in gregariousness among individuals can be a byproduct of the Bejder et al. (1998) test for preferred/avoided companionship (Section 4.9; Whitehead et al. 2005). A suitable test statistic is the standard deviation of typical group sizes among individuals:

$$s = SD\left[\frac{\sum_{k}\left[x(k, I) \sum_{J} x(k, J) \right]}{\sum_{k} x(k, I)} \right] \quad (3)$$

where $x(k, I) = 1$ if individual I is a member of group k, and $x(k, I) = 0$ if it is not. The quantity s is calculated for the real data as well as for data sets constructed by permuting the real data in such a way that the number of individuals in each group and the number of groups containing each individual are held constant (Section 4.9; Bejder et al. 1998). Significantly large values of s indicate that some animals are consistently found in particularly large groups and others in particularly small groups, and so there are differences in gregariousness (Section 2.4).

If the data consist of rates of interaction or associations not defined using groups, then another approach to testing for differences between mean rates of interaction or association among individuals is needed. We can simply use a one-way analysis of variance (Sokal & Rohlf 1994, pp. 207–271) or its nonparametric equivalent, the Kruskal-Wallis test (Sokal & Rohlf 1994, pp. 423–427), on interaction rates, or typical group sizes, of each individual in each sampling period, but these are theoretically invalid because the data for the different individuals are not independent (Section 2.4).

Table 4.3 Numbers of Interactions among Individuals within a Sampling Period (above), with Identities Randomized (below) for Permutation Test for Individual Differences in Interaction Rates

	Real data for sampling period							
	JOE	SAL	FRED	BOB	SUE	CON	ART	BILL
JOE	0	8	4	2	6	1	0	3
SAL	0	0	1	7	3	2	1	1
FRED	9	1	0	0	5	0	2	1
BOB	0	0	0	0	1	1	0	0
SUE	1	4	1	1	0	0	6	1
CON	0	3	3	0	0	0	9	1
ART	0	0	1	0	0	0	0	1
BILL	7	5	9	6	2	2	1	0

↓

	Permuted data for sampling period							
	ART	BILL	SAL	FRED	CON	BOB	JOE	SUE
ART	0	8	4	2	6	1	0	3
BILL	0	0	1	7	3	2	1	1
SAL	9	1	0	0	5	0	2	1
FRED	0	0	0	0	1	1	0	0
CON	1	4	1	1	0	0	6	1
BOB	0	3	3	0	0	0	9	1
JOE	0	0	1	0	0	0	0	1
SUE	7	5	9	6	2	2	1	0

An alternative, suggested by Whitehead et al. (2005), is to construct random association or interaction data by permuting the identities of the individuals in each sampling period, while retaining the numeric structure of the data (Table 4.3), as in the Mantel test (Mantel 1967; Section 2.4). All individuals could be permuted at each sampling period, but so that demographic effects (movement in and out of the study area, recruitment, mortality) do not produce significant results, I suggest only permuting the identities of individuals actually identified in each sampling period. A test statistic (such as the SD of the mean number of associates, or interactions, of each individual over sampling periods during which it was identified) is calculated for the real data and compared with its distribution over all random data sets (Section 2.4). This test controls for the possibility that individuals may have generally associated or interacted more in some sampling periods than in others.

As an example of how it works, consider the results for a data set on bottlenose whales (*Hyperoodon ampullatus*), summarized in Table 1.3 for an earlier published analysis, containing 160 identified individuals. If association is defined as "identified within 15 minutes of one another" and sampling period is 1 day, the SD, among individuals, of the mean number of associates of each individual (over days on which each individual was identified) was 1.29 individuals. The mean of this test statistic for 1,000 random data sets (produced by permuting the identities of the individuals identified on each day) was 1.40 individuals. In 58 cases, the statistic for the random data was less than the real one, and so the alternative hypothesis that individuals differ in their gregariousness (producing an abnormally high real SD) is not accepted at $P = 0.94$.

These techniques of examining differences in overall gregariousness and interaction rates have some of the benefits of Bejder et al.'s (1998) test for preferred companions (Section 4.9). P values can be calculated for each individual (the proportion of permutations with typical group size, mean number of associates or mean interaction rate for that individual less than the real typical group size, mean number of associates or mean interaction rate for that individual) to identify individuals with significantly high or low gregariousness or interaction rates. Another potential extension is to examine variations between classes of individuals, so that the tested hypothesis is something like "Do females differ in the number of males with which they associate?" I have not seen this done.

4.4 Rates of Interaction

A fundamental method by which to describe the content of a relationship between two individuals is to use their rates of interaction (Altmann & Altmann 1977; Michener 1980), such as the number of grooming instances per sampling period. Some of these relationship measures will be symmetric (e.g., the rate at which A and B touch), and others will not be (e.g., rate of agonistic interactions in which A dominates B). With rare events, it may be appropriate to use 1:0 instances, rather than rates, of interaction (e.g., did A and B ever fight during the study?).

Table 4.4 shows an example from Perry (1996) of a matrix of grooming rates. This is an asymmetric measure, and it is clear that individuals differed greatly in the rates at which they groomed and were groomed.

An important consideration when using rates of interaction as a relationship measure is effort. Unless all individuals were observed all of the time, we should not simply tally observed numbers of interactions for all dyads (Lehner 1998, p. 201). Interaction data are generally recorded during focal individual (Section 3.7; Table 3.4), or occasionally focal group, follows. With focal individual follows (Section 3.7; Altmann 1974), the measure of effort for a dyad is usually the total time, or

Table 4.4 Rates of Female–Female Grooming in Capuchin Monkeys (*Cebus capucinus*), in Seconds Grooming/Hour

Actor	Recipient						
	A	S	N	D	W	T	Mean
A	–	5.8	3.5	2.1	2.3	0.04	2.7
S	41.6	–	28.6	18.1	9.0	7.4	20.9
N	10.3	25.5	–	9.6	9.9	4.3	11.9
D	23.3	9.3	10.5	–	13.4	6.9	12.7
W	21.2	15.2	14.6	25.1	–	10.4	17.3
T	2.5	2.9	3.7	3.6	5.3	–	3.6
Mean	19.8	11.7	12.2	11.7	8	5.8	11.5

From Perry (1996).

number of sampling intervals, in which either of the two individuals was focal. For focal group follows, the effort should either be the amount of time spent following groups in which either individual is present, or that spent following groups in which both are present or available to interact. These give distinctively different interaction rates, the former being an estimate of the absolute interaction rate of the dyad over time, the latter only over those times in which the dyad was in a position to interact [assuming "group" is defined as individuals in a position to interact (Section 3.4; see also Michener 1980)]. In any case, care must be taken that there are no substantial biases (e.g., rates of recording interactions that do occur among followed individuals do not vary among individuals or classes of individuals).

The precision of interaction rates can be estimated using bootstrap or jackknife techniques (Section 2.3) in which sampling periods are resampled with replacement (bootstrap) or omitted in turn (jackknife). These assume that the data obtained in different sampling periods are independent.

If the interactions themselves are assumed to be independent (and so Poisson distributed), and n_{AB} interactions were observed in e_{AB} units of effort, then the interaction rate and its standard error can be estimated from

$$I_{AB} = \frac{n_{AB}}{e_{AB}} \tag{4}$$

$$SE(I_{AB}) = \frac{\sqrt{n_{AB}}}{e_{AB}} \tag{5}$$

Using these formulas, we can calculate the coefficient of variation (CV = SE/mean) of the interaction rate. These are presented for different numbers of observed interactions in Table 3.14. In cases in which interactions are assumed to be independent, Poisson-derived 95% confidence intervals for interaction rates are given by

$$95\%\,\mathrm{CI}(I_{\mathrm{AB}}) = \left\{ \frac{\chi^2_{2n_{\mathrm{AB}},0.025}}{2e_{\mathrm{AB}}}, \frac{\chi^2_{2n_{\mathrm{AB}},0.975}}{2e_{\mathrm{AB}}} \right\}$$

where $\chi^2_{x,p}$ is the value at which the cumulative probability of the chi-squared distribution with x degrees of freedom equals P (commonly presented in statistical tables).

4.5 Association Indices

Following interaction rates, *association indices* form the second major class of relationship measures. In species in which interactions are hard to observe or relationships are best expressed through associations rather than interactions (Section 3.7), association indices become the fundamental building blocks for describing social structure. For instance, in my studies of whale societies, an invariant step has been the calculation of association indices among identified individuals. Calculating association indices was a fundamental driver for the initial development of the software, SOCPROG.

If all animals and their associations (Section 3.3) are visible during all of the observation time, then the proportion of time each dyad spends associated can be estimated directly, or counts of "joint occurrences" can be used as a proportional stand in (Table 4.5). Usually, however, we need to standardize the number of observations of association by some measure of effort. Unfortunately, converting records of associations into relationship measures is not quite as straightforward as calculating rates of interaction. A variety of association indices is available (Cairns & Schwager 1987; Whitehead & Dufault 1999; Table 4.5), most of which were developed for ecological applications and then incorporated, without too much consideration, into social analyses. Almost all association indices estimate the proportion of time that a pair of individuals spends in association, and so they are symmetric (the association index of A with B equals that of B with A) and range between zero and one. Exceptions are the simple "joint occurrences" and Cole's (1949) index, which can be negative, and seem to have few advantages over other available indices. Depending on the method by which the data were collected, one or

Table 4.5 Commonly Used Association Indices

Index	Formula	Comments
"Joint occurrences"	x	Effort not controlled; does not estimate proportion of time together
"Simple ratio"	$\dfrac{x}{x+y_{AB}+y_A+y_B}$	Unbiased if assumptions hold (Ginsberg & Young 1992)
"Half-weight" (also "Dice's," "Sorensen's," "coherence")	$\dfrac{x}{x+y_{AB}+\frac{1}{2}(y_A+y_B)}$	Most commonly used; less biased when individuals are more likely to be identified when not associated or not all associates identified; monotonic function of twice-weight
"Twice-weight"	$\dfrac{x}{x+2y_{AB}+y_A+y_B}$	Less biased when individuals are more likely to be identified when associated; monotonic function of half-weight
"Square root"	$\dfrac{x}{\sqrt{(x+y_{AB}+y_A)(x+y_{AB}+y_B)}}$	Based on flawed probability model (Cairns & Schwager 1987)
"Social affinity"	$\dfrac{x}{\min\{(x+y_{AB}+y_A)\cdot(x+y_{AB}+y_B)\}}$	May be useful when individuals differ considerably in their identifiability
"Both identified"	$\dfrac{x}{x+y_{AB}}$	Controls for cooccurrence

x, number of sampling periods with A and B observed associated; y_A, number of sampling periods with just A identified; y_B, number of sampling periods with just B identified; y_{AB}, number of sampling periods with A and B identified but not associated.
From Cairns and Schwager (1987), Whitehead and Dufault (1999), and Christal and Whitehead (2001).

other of these indices may be more appropriate, as discussed by Cairns and Schwager (1987), Ginsberg and Young (1992), and later in this section. Here are some assumptions for an ideal data set:

1. Recorded association is a symmetric 1:0 measure of whether the members of a dyad are or are not associated in a sampling period.
2. Recorded associations are accurate.
3. If one individual is identified in a sampling period, then all its associates are identified.
4. Members of a dyad are equally likely to be identified whether they are associated or not associated.

If these assumptions hold, then the "simple ratio" index—simply the ratio of the number of sampling periods in which two individuals were recorded as associated divided by the number of sampling periods in which at least one of them was identified (formula in Table 4.5)—is

Table 4.6 Expected Values of Association Indices under Various Scenarios of Association and Identification

Identification rate	Association rate	"Simple ratio"	"Half-weight"	"Twice-weight"	"Squareroot"	"Social affinity"	"Both identified"
				Association indices:			
			Individuals identified independently[a]				
0.7	0.75	**0.750**	0.796	0.661	0.796	0.796	0.848
0.7	0.25	**0.250**	0.302	0.178	0.302	0.302	0.382
0.1	0.75	**0.750**	0.851	0.740	0.851	0.851	0.983
0.1	0.25	**0.250**	0.388	0.241	0.388	0.388	0.864
			Pairs identified independently[b]				
0.7	0.75	0.698	**0.750**	0.600	**0.750**	**0.750**	0.811
0.7	0.25	0.204	**0.250**	0.143	**0.250**	**0.250**	0.323
0.1	0.75	0.612	**0.750**	0.600	**0.750**	**0.750**	0.968
0.1	0.25	0.149	**0.250**	0.143	**0.250**	**0.250**	0.769
			Identified at half the rate when alone[c]				
0.7	0.75	0.784	0.857	**0.750**	0.857	0.857	0.945
0.7	0.25	0.288	0.400	**0.250**	0.400	0.400	0.656
0.1	0.75	0.755	0.857	**0.750**	0.857	0.857	0.992
0.1	0.25	0.255	0.400	**0.250**	0.400	0.400	0.930
			Fifty percent of associates not identified[d]				
0.7	0.75	0.577	0.674	0.508	0.674	0.674	**0.811**
0.7	0.25	0.192	**0.241**	0.137	**0.241**	**0.241**	0.323
0.1	0.75	0.395	**0.561**	0.390	**0.561**	**0.561**	0.968
0.1	0.25	0.132	**0.225**	0.127	**0.225**	**0.225**	0.769
			Individual B identified 25% time present[e]				
0.7	0.75	0.560	0.687	0.523	**0.699**	0.846	0.887
0.7	0.25	0.124	0.196	0.108	**0.223**	0.378	0.465
0.1	0.75	0.547	0.703	0.542	**0.715**	0.856	0.983
0.1	0.25	0.118	0.208	0.116	**0.237**	0.397	0.868
			Individual B dies 25% way through study[f]				
0.7	0.75	0.227	0.358	0.218	0.429	**0.796**	0.848
0.7	0.25	0.076	0.126	0.067	0.155	**0.302**	0.382
0.1	0.75	0.291	0.449	0.289	0.509	**0.851**	0.983
0.1	0.25	0.097	0.174	0.095	0.209	**0.388**	0.864

Bold values indicate associate indices that match the true association rate (true association index) most closely. The identification rate is the probability that an individual is identified in a sampling period.

[a] Individuals have a fixed probability of being identified primarily (the identification rate); if they are identified, all of their associates are secondarily identified.

[b] Groups of associates have a fixed probability of being identified (the identification rate); if a group is identified, all of its members are identified.

[c] When associated, individuals have a fixed probability of being identified (the identification rate); when not associated, this probability is halved.

[d] Individuals have a fixed probability of being primarily identified (the identification rate); if they are identified, then 50% of their associates are secondarily identified.

[e] Individual A has a fixed probability of being primarily identified (the identification rate); individual B is primarily identified at 25% of this rate; if A is primarily identified, then B is secondarily identified with probability 0.25; if B is primarily identified, then A is secondarily identified with probability 1.0.

[f] Individuals have a fixed probability of being identified primarily (the identification rate), although this drops to 0.0 for individual B in the last 75% of the study, as does the association rate between A and B

an unbiased estimate of the proportion of time they spend together (Ginsberg & Young 1992; Table 4.6). This is why Ginsberg and Young (1992) recommend the general use of the "simple ratio" index, and it has been quite commonly used as a measure of relationship in the years since their paper was published.

If the assumptions do not hold, then the "simple ratio" index will probably be biased (Table 4.6), and other indices may be more appropriate. In the following subsections, I consider the effects of failures in each of the foregoing assumptions, with some possible remedies. I take the assumptions in reverse order, and then conclude with how to estimate the precision of association indices. Some general recommendations on the choice of an association index are summarized in Box 4.1. Table 4.6 shows the expected biases of different association indices in different circumstances. The discussion of "special" relationships in Section 4.9 includes methods of testing whether association indices are particularly large or small, as well as correcting association indices for the gregariousness of individuals.

4.5.1: Identification and Association Linked. Quite often, members of a dyad will be more or less likely to be identified when associated than when not associated. If association is defined based on group membership, the former will result when large groups are more easily detected than small ones are, whereas the latter will be a consequence of a situation in which each group has the same probability of being identified, so that when the pair is separated into two groups, there is a greater probability of identifying at least one member (Cairns & Schwager 1987). Cairns and Schwager (1987) show that the "twice-weight" index is less biased than the "simple ratio" in the case in which pairs are more likely to be identified together, and the "half-weight" index (or the "square root" or "social affinity" index) is better when pairs are more likely to be identified when apart, and is in fact unbiased if each group has the same probability of being identified (Table 4.6). This makes sense because, using the terminology of Table 4.5, in the first case, the measured x_{AB} (the number of times A and B are identified as being associated) will be too high and the "twice-weight" index reduces its significance, whereas in the reverse situation, x_{AB} pwill be too low and the "half-weight" index inflates its significance. Cairns and Schwager (1987) develop almost unbiased maximum likelihood indices, which theoretically work very well. However, this method requires a specific and realistic model of the data collection methodology, is technically relatively complex, and has rarely been used in practice.

4.5.2: All Associates Not Identified. I can envisage two principal situations in which all the associates of an identified individual are not themselves identified. With cryptic species, an associate of an identified individual may be hidden (in foliage, beneath the water, or some other way) or visible but not seen well enough to be identified. Second, if the sampling period is relatively long and an individual is only viewed for a small proportion of it, it may have other associates during parts of the sampling period when it was not being viewed.

In these cases, y_A and y_B (counts of periods in which only one individual is identified) will be biased upward and x (periods with individuals associated) downward, lowering the association index. This means that, as with the case when individuals are identified more when apart (see previous discussion), the "half-weight," "square root," and "social affinity" indices generally correct at least somewhat for the bias in the simple ratio index (Table 4.6). Bias caused by missing associates within long sampling periods can also be lessened by reducing the length of the sampling period to less than the usual time for disassociation (see Section 5.5).

In some situations, individuals may differ considerably in their identifiability, and then the usual indices will be biased downward because if A is the much more identifiable individual, then y_A (only A identified) will be much larger than the other elements of the formulas. The "social affinity" index and especially the "square root" index, which operate more from the perspective of the least observable individual (Table 4.5), remove much of this bias (Table 4.6).

4.5.3: Recorded Associations Are Inaccurate. If there are errors in recording associates but they are not biased toward recording proportionally more or fewer associates than exist, then the "simple ratio" index should remain approximately unbiased overall, although, generally, low association indices will be increased and high ones reduced. Biases toward recording more associates will increase association indices and suggest the use of the "twice-weight" index to reduce the effect. Similarly, the "half-weight" index can be used to reduce the effects of systematically underrecording associates (see previous subsection).

4.5.4: Asymmetric or Non-1:0 Association Measures. Association indices for asymmetric (if A is associated with B, then B is not necessarily associated with A) or non-1:0 association measures have not been much considered (Section 3.3). The simplest way to deal with an asymmetric measure is probably just to use it to produce a new symmetric measure. For instance, A and B might be considered associated if either A is B's nearest neighbor or B is A's nearest neighbor, giving a symmetric index that can be used to produce association indices (as in Table 4.5). Little important information will be lost in most cases by using this procedure. The true proportion of time that individuals are associated, which is the target of most association indices, is naturally symmetric.

With continuous, or other non-1:0, measures of association between each dyad in each sampling interval, it would seem reasonable to use the mean of the measure over sampling periods as an association index. The major problem then is what to do with sampling periods in which neither, or only one, of the individuals was identified. There are various possible approaches that can be tailored to the particular situation, usually involving omitting sampling periods in which neither individual was identified; see Whitehead and Arnbom (1987) and Perry (1996) for some possibilities.

4.5.5: The Precision of Association Indices. Like other measures of relationship, the precision of association indices can be estimated using bootstrap or jackknife techniques (sampling periods are resampled with replacement for bootstrap or omitted in turn for jackknife; Section 2.3). Alternatively, if x is the number of sampling periods in which A and B were observed associated and α is the calculated association index (as in Table 4.5), then, using the binomial distribution, we can estimate the standard error of α from

$$\mathrm{SE}(\alpha) = \alpha\sqrt{\frac{1-\alpha}{x}} \tag{6}$$

From this, the coefficient of variation (CV = SE/mean) of association indices can be estimated. These are tabulated for different levels of α and x in Table 3.14. All of these methods assume that the associations observed in different sampling periods are independent. The bootstrap, although it may take some computational time with large data sets, is probably the best of the available techniques, although in practice, standard errors from the binomial approximation [Equation (6)] and bootstrap are generally in close agreement (Whitehead In press-a).

Both the binomial approximation [Equation (6)] and bootstrap methods of estimating the precision of association indices possess the drawback that if the estimated association index is either $\alpha = 0.0$ (i.e., $x = 0$; never seen associated) or $\alpha = 1.0$ (always seen associated), then the estimated standard error is exactly 0.0. However, there may be considerable uncertainty about an estimate of either zero or complete association. Better than using the CV in such circumstances is to construct confidence intervals using methods such as Wilson's (1927) score for binomial proportions (here given for a 95% confidence interval):

$$95\% \,\mathrm{CI}(\alpha) \;=\; \frac{\alpha + z_{0.975}^2/2d \pm z_{0.975}\sqrt{[\alpha(1-\alpha) + z_{0.975}^2/4d]/d}}{1 + z_{0.975}^2/d}$$

where $z_{0.975}$ is the 97.5th percentile of the normal distribution and d is the denominator of the association index (as given in Table 4.5, or x/α, unless $\alpha = 0$).

BOX 4.1 *Choosing an Association Index: Recommendations*

When choosing an association index to measure the relationship between two individuals, there are several issues the social analyst should consider.

Do you wish to correct for bias? It is not easy to predict which index will best correct for bias so that association indices more accurately reflect the proportion of time that a dyad spends together. Ginsberg and Young (1992) argue that, because of the difficulty and arbitrariness of making such corrections, it is best just to use the "simple ratio" index and discuss how biases might affect the results observed. Association indices are actually used rather rarely as estimates of the proportion of time that individuals spend together, although if this is the case, bias is obviously to be avoided as far as possible. More frequently, association indices are compared among dyads or within populations (e.g., Sections 4.9, 5.2, and 5.3). If the bias is similar for all dyads, then for within-population comparisons, it is of little concern. For between-population comparisons, the bias is unlikely to be similar, so it may be worth choosing an index to reduce bias if the indices are to be compared with those from a different population.

Should the index include or exclude demographic effects? Individuals may not be associated because they choose not to be or because of what I call demographic effects: Two animals cannot associate during a sampling period if one has not been born, has died, or has temporarily or permanently emigrated from the study area. An individual that emigrated part of the way through a study will generally have a low value on most standard association indices with those that stayed on, so that demographic effects are included in the value of the index (see bottom section of Table 4.6). This may be what the researcher desires if she is using the association index to measure the potential for food competition. In contrast, if she is examining how social relationships correlate with kinship, it makes sense when constructing association indices to consider only times during which a dyad could have associated. The "social affinity" index (Table 4.5) largely excludes demographic effects and is useful in such circumstances.

Putting these considerations together, I generally recommend using the simple "simple ratio" index, except in the following cases:

- When there is a clear source of bias (from individuals being more or less identifiable when together or apart, not all associates being identified, or some other cause) AND the researcher wishes to use the absolute values of the index OR to make interpopulation comparisons. Then it may be useful to use the "half-weight" index, the "twice-weight" index, a maximum-likelihood index (Cairns & Schwager 1987), or some other association index.
- When the researcher wishes to EXCLUDE demographic effects from the index. Then she should use the "social affinity" index or something similar.

Additional recommendations are to use a short sampling period to reduce bias and, if uncertain about which is the best association index, to try two or more association indices and then see if the conclusions of the analysis are changed.

4.6 Temporal Patterning of Interactions/Associations

Rates of interaction and association indices provide quite straightforward measures of the content and quality of relationship, but the third of Hinde's (1976) features of a relationship—temporal patterning—is less easily quantified. However, it should not be ignored. A relationship between a pair of humans changes over many temporal scales, including minutes, hours, weeks, months, and years, as do those of nonhumans. de Waal (1998) vividly describes the very dynamic nature of the relationship between Yeroen and Luit, two male chimpanzees (*Pan troglodytes*) at the Arnhem Zoo, over all these scales. Such a written history of a relationship will only be possible and desirable in very few cases. Rarely have two animals been watched as closely and carefully as Yeroen and Luit, and the description of their interactions and associations fills the large part of a book. We need techniques for abstracting information on temporal patterning from much more sparse data as well as for summarizing it.

Temporal patterning can be seen from three principal conceptual perspectives: developmental, cyclical, and fission–fusion. Each has a set of corresponding statistical methods, which could be characterized as regressive, spectral, and autoregressive analyses. The last two of these usually assume "stationary" models (Dunstan 1993), so that the long-term nature of the relationship does not change, whereas with a developmental model, long-term change is the essence.

The archetypal developing relationship is between parent and offspring. This can be described using a plot of interaction or association measures with offspring age. An example is shown in Fig. 4.1, which is averaged over five infant dolphins, whereas in Fig. 4.2 the developing relationship is between an infant and her older brother. Usually such data are simply plotted against age or time as in these examples. It is possible, however, to fit regression and other models to such data (Sokal & Rohlf 1994, pp. 451–545, 609–681).

Second, a relationship's quality might vary in some cyclical manner, in which case association or interaction measures could be analyzed using spectral analysis [for an introduction, see Dunstan (1993)]. I have yet to see this done, and am not sure such models will be appropriate except in a very few cases. Behavior does often vary cyclically, but almost always because there is an underlying periodic force, either environmental, such as diurnal or seasonal cycles, or physiological, such as reproductive cycles. In these cases, a spectral analysis may describe variation in behavior quite well, but relating the measure of behavior to the forcing factor directly (whether time of day, season, or hormonal level) will be much more informative. Spectral analysis may perform usefully in another case: when the underlying model is autoregressive, so that the value of the variable in one sampling unit affects its value in subsequent ones, but in such cases, a true autoregressive model is preferable and, in some ways, simpler.

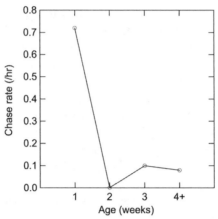

FIGURE 4.1 Rates at which mother dolphins (*Tursiops* spp.) chase their infants as a function of infant age. (Redrawn from Mann & Smuts 1998).

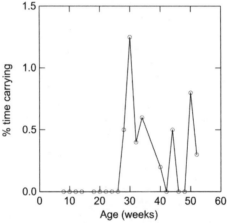

FIGURE 4.2 Development of a relationship between a captive infant gibbon (*Hylobates pileatus*) and her older brother: the percentage of time the infant was carried by her brother as a function of infant age. (Redrawn from Geissmann & Braendle 1997.)

Autoregressive systems are particularly important in social analysis because they can be used to analyze and model the classic fission–fusion social systems of many higher vertebrates. *Fission–fusion* has been much written about but less often defined. A recent definition of a fission–fusion society captures the essence: "a society consisting of casual groups of variable size and composition, which form, break-up and reform at frequent intervals" (Conradt & Roper 2005). In defining a relationship within a fission–fusion society, the challenge is to characterize this dynamism.

One approach is to divide the study into time periods (each containing several of what I call sampling periods; Section 3.9), calculate rates of interaction (Section 4.4) or association indices (Section 4.5) for each period, and then compare them. They can be plotted against time, or the standard deviation or coefficient of variation of the interaction rates or association indices among time periods can be used as measures of temporal variability in a relationship (Whitehead 1997). Standard deviations and coefficients of variation are confounded with measurement error, however, and depend on the time periods chosen. There are better alternatives that use an approach based on autocorrelation.

Suppose that we have measured the interaction rates of a dyad over several time periods. Then, for any time lag τ longer the interval between the periods (e.g., 1 minute, hour, or day), we can characterize the change in their interaction rate using the autocorrelation function. The autocorrelation of lag τ is simply the correlation coefficient between the interaction rates at time periods $\{t_1, \dots, t_T\}$, and time periods $\{t_1 + \tau, \dots, t_T + \tau\}$, where interaction rates I are assumed to be available for all these time periods. Thus,

$$\rho(\tau) = r(I(t_1, \dots, t_r), I(t_1 + \tau, \dots, t_T + \tau))$$

A high autocorrelation (ρ close to 1.0) indicates that interaction rates change little over time scales of τ, a value near 0 indicates little holdover in rates of interaction over such time periods, and a low autocorrelation ($\rho < 0.0$) indicates that a high interaction rate at any time is likely to be followed by a relatively low one τ units later or vice versa. The autocorrelation with lag τ is then plotted against τ in a display known as a correlogram (Dunstan 1993), which indicates time scales of changes in interactions for the dyad. I do not know of any such analyses for interaction rates of a dyad, but they are certainly feasible if sufficient data are available over a range of time scales.

The same approach can be used for 1:0 measures such as associations. If there are N pairs of sampling periods τ units apart, and if the dyad is associated during n of the sampling periods in each series, and is associated in the corresponding periods of both series during $m(\tau)$ pairs of periods, then the autocorrelation is

$$\rho(\tau) = \frac{m(\tau) - n^2/N}{n - n^2/N} \qquad (7)$$

This is closely related to a simpler measure that I call the *lagged association rate* (Whitehead 1995):

$$g(\tau) = m(\tau)/n \tag{8}$$

The lagged association rate is simply the probability of association τ time units after a previous association. The lagged association rate $g(\tau)$ is 1.0 if an associated pair is always associated τ time units later [equivalent to $\rho(\tau) = 1$], and $g(\tau) = 0.0$ if associations always disband and do not reform within τ time units [equivalent to $\rho(\tau) = -1$]; if there is no relationship between associations τ time units apart, then $g(\tau) = n/N$ [equivalent to $\rho(\tau) = 0$]. I call $g(\tau) = n/N$ the *null association rate* (Whitehead 1995). I have assumed so far that n, the number of recorded associations, is the same for both series of time periods. This is not normally the case, although they should be similar if the relationship is stationary in the statistical sense (i.e., not changing systematically with time). For the mathematically estimated lagged association rate to conform to the informal definition ("rate of association τ time units after a previous association"), then n should refer to the number of associations in the earlier series.

The lagged association rate is usually plotted against lag (τ) for a range of values of τ. An example is shown in Fig. 4.3, which presents the lagged and null association rates, for two female sperm whales (*Physeter macrocephalus*). These whales seems to have had periods of stronger association lasting about 1 week, after which their association rate fell to near their long-term mean (as indicated by the null association rate).

Unless there are considerable data for a dyad, lagged association rates, or autocorrelation analyses of interaction rates, will have little validity for characterizing the temporal patterning of a particular relationship. However, these techniques can be generalized to describe temporal patterns of relationship within an entire population or community, as well as relationships within and between classes of individual. They then become powerful. Such generalizations, as well as a number of extensions and variations of the method of using lagged association rates (including the use of the jackknife technique to obtain confidence intervals about lagged association rates, the use of "standardized" rates in situations when not all interactions or associates are recorded, and model fitting), are described in Section 5.5. Although sufficient data will rarely be available, the fitting of mathematical models representing different types of relationship is possible for dyadic association data; the methods summarized in Section 5.5 for populations, communities, and

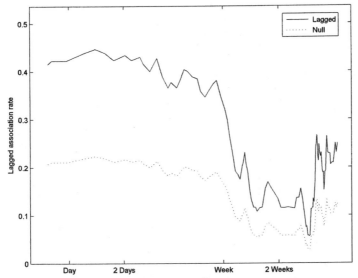

FIGURE 4.3 Lagged and null association rates for a pair of female sperm whales (*Physeter macrocephalus*), ID#3703 and ID#3708, both members of "social unit T," observed off the Galápagos Islands in 1998 and 1999. Sampling periods are 1 hour long (83 sampling periods with at least one of the individuals identified), and association is defined by diving within 5 minutes. Lagged association and null association rates are plotted using a moving average of 20 recorded associations.

classes can usually be used for dyads without modification (standardized lagged association rates are an exception).

4.7 Relative Relationships: Multivariate Description of Relationships

The methods described in the previous three sections (interaction rates, association indices, and lagged rates) are absolute measures of a relationship: They can be calculated for a dyad and used to describe the relationship of its members without reference to other members of the community. We can say that A and B have "an association index of 0.4, and when they are together, they have affiliative interactions at a rate of 0.3/hour, agonistic interactions at a rate of 0.1/hour, A grooms B 8% of the time, and B grooms A 5% of the time." We might add that "the lagged association rate between A and B falls to the null association rate after about 3 hours." Such statements describe the A-B relationship in absolute terms.

It is also possible and often useful, however, to consider the relative strength of a relationship with reference to either all of the relationships among members of the community or just those that involve one or other of the members of the dyad under concern. Thus, we can describe the relationship between A and B as "having an association index greater than 75% of those among dyads of their community, with an affiliation rate 150% of the community mean and a rate of agonism at 80% of the community mean."

When several relationship measures are available, as in the previous example, powerful and informative methods are available (Whitehead 1997). For instance, each relationship measure can be represented by an axis in multivariate space and each dyad by a point. Then the position of a particular dyad can be assessed with respect to the overall distribution of points and the axes.

As an example, Fig. 4.4 displays four measures of the relationship between a pair of chickadees at a feeder who later mated, BJAO and SORA, relative to the relationship measures of the other dyads in the data set. From the comparative perspective, this dyad was particularly likely to be nearest associates but not to arrive at the feeder together or to be censused together in the same hour. The other two pairs that later mated were also frequently nearest associates, but, unlike BJAO and SORA, often arrived together and were censused in the same hour. In addition, BJAO and SORA showed relatively little asymmetry in their agonistic interactions (Fig. 4.4).

If relationship measures are correlated, then dimensionality can be reduced using techniques such as principal components analysis (Whitehead 1997; Section 2.6). Other techniques of analyzing several relationship measures discussed in Section 5.6 may also be useful when considering a particular relationship.

4.8 Types of Relationships

Although relationships vary enormously, they clearly form categories. Categorizing relationships is not only a useful step by which an analyst may summarize social structure, the animals may do this themselves, treating one set of individuals in one way and another set in another. Thus, in common, as well as ethological, vocabulary we give names to certain types of relationship. Some are nouns, such as "bonds" and "acquaintances," whereas others are adjectives applied to the word "relationship": for example, dependent relationships, dominance relationships, asymmetric relationships, and reciprocal relationships. In this section and those that follow (Sections 4.9 and 4.10), I consider some of these terms and whether there are ways to define them that allow statistical or experimental assignation of a particular relationship into the type using interaction or association data.

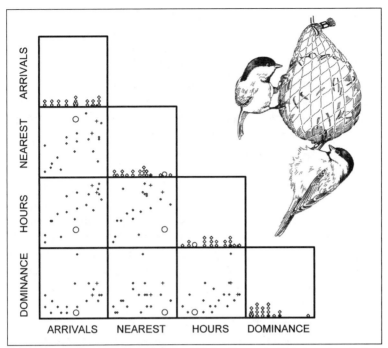

FIGURE 4.4 Four relationship measures for seven chickadees (*Parus atricapillus*) observed at a feeder: arrivals at feeder within 1 minute; nearest associates; within same hourly census; and difference in dominance measure. Each symbol represents one dyad: o, dyad BJAO–SORA; +, other dyads that later formed mated pairs; •, nonpaired dyads. (Data from Ficken et al. 1981.) (Illustration copyright Emese Kazár.)

There is an important challenge in discriminating types of relationship. Although it may be possible to derive a statistical hypothesis test of whether a particular relationship fits into a type—whether a pair is "bonded," for instance—the outcome of the test will depend on both the "strength" of the relationship—what we are interested in—and the amount of data available (and so the power of the test), as well as chance and the chosen significance level. The most responsible procedure is to focus on the strength of the effect—how well the data indicate that the relationship falls into a particular type—and to use the results of a statistical significance test to indicate our confidence in this assignment (Section 2.1). Using this procedure, we represent each type by a measure of effect size, a "strength," "bondedness," "asymmetry," "dominance," or "dependence." To assign "friends," "bonds," asymmetric relationships," "dominant relationships," and "dependent relationships," we must choose a cutoff on this scale. Usually this is rather arbitrary, so I tend to focus on the measure of effect size rather than ascribing particular relationships to types, saying "the relationship between A and B has an asymmetry of 3.2" rather than "A and B have an asymmetric relationship."

In this section, I consider types of relationship that can be derived principally from interaction data. In Sections 4.9 and 4.10, the emphasis is on association data. Some attributes of a relationship can be assessed using the information for the dyad in isolation. An example of such an absolute measure is the symmetry of interaction rates. Other attributes are relative and can only be assessed by comparison with the other relationships present in the community.

4.8.1: High and Low Interaction Rates. The most obvious way to characterize a relationship is through interaction rates. With a matrix of interaction rates, such as those in Table 4.4, some dyadic rates are always relatively high or low. Can these be explained just by sampling? By overall differences in interaction among individuals? Or by interaction rates specific to a dyad? Thus, we may wish to examine a number of factors that can affect interaction rates.

If we have asymmetric interaction rates for pairs within a community $I_{AB}(t)$ estimated for a number of sampling intervals t, we can express these as an ANOVA-type general linear model (Kirk 1995, pp. 219–220) with several possible factors: an overall mean interaction rate (μ), a sampling period effect (ν_t), an actor effect (α_A), a receiver effect (β_B), a dyadic effect (γ_{AB}) and the error (ε). Various models represent different hypotheses about relationships within the community:

$I_{AB}(t) = \mu + \nu_t + \varepsilon$: There are no individual or dyadic effects,
    all relationships are effectively the same, but their intensity
    might vary between sampling periods.

$I_{AB}(t) = \mu + \nu_t + \alpha_A + \varepsilon$: Individual-specific actor rates are the
    principal influence on relationships.

$I_{AB}(t) = \mu + \nu_t + \beta_B + \varepsilon$: Individual-specific receiver rates are
    the principal influence on relationships.

$I_{AB}(t) = \mu + \nu_t + \alpha_A + \beta_B + \varepsilon$: Individual-specific actor and
    receiver rates are the principal influences on relationships.

$I_{AB}(t) = \mu + \nu_t + \gamma_{AB} + \varepsilon$: Dyads have characteristic interaction
    rates.

The fit of the data to these models can be assessed using AIC methods, F tests, or likelihood-ratio tests (Section 2.8; Burnham & Anderson 2002), and the relative importance of the different factors could be indicated by Akaike weights (Burnham & Anderson 2002, pp. 167–169).

With symmetric interactions, the range of models is narrowed, with α_A becoming an individual (rather than an actor or receiver) effect:

$I_{AB}(t) = \mu + v_t + \varepsilon$: There are no individual or dyadic effects, all relationships are effectively the same, but they might vary between sampling periods.

$I_{AB}(t) = \mu + v_t + \alpha_A + \alpha_B + \varepsilon$: Individual-specific interaction rates are the principal influences on relationships.

$I_{AB}(t) = \mu + v_t + \gamma_{AB} + \varepsilon$: Dyads have characteristic interaction rates.

The general linear model assumes that errors (ε) are normally distributed. Generalized linear models (Dobson 2001) allow nonnormal error structures and link functions so that the left-hand side of the model equation is some function of $I_{AB}(t)$, perhaps a log. An important type of generalized linear model is the log-linear model (Sokal & Rohlf 1994, pp. 743–760) for categorical data, an extension of the "chi-squared test." This can be used in the case in which all interactions can be assumed to be independent, and so we do not need to consider sampling intervals. The model, then, with asymmetric interactions and dyadic rates is

$$\log_e(I_{AB}) = \log_e(E_{AB}) + \mu + \gamma_{AB} + \varepsilon$$

where, in this case, I_{AB} is the number of interactions between A and B and E_{AB} is the effort spent observing this dyad. The importance of the different model terms can be assessed in similar ways as before. If effort is equal for all dyads, then the $\log_e(E_{AB})$ term can be omitted, and the model reverts to a more tractable two-way G test or chi-squared test (Sokal & Rohlf 1994, pp. 724–743), with the important modification that diagonal terms (animals interacting with themselves) are omitted.

As an example of this approach, several models are fitted to the grooming rates of a community of eight chimpanzees measured in two time periods (Table 4.7). The best-fitting model of those used, indicated by that with the lowest AIC (in Table 4.8), contains just a groomee effect: Some animals are groomed more than others (mean rates of being groomed are 9.5, 18.0, 9.5, 12.5, 43.0, 9.0, 6.5, and 41.5, respectively, for the eight individuals). The identity of the groomer, dyadic effects, and differences between the time periods seem to be relatively unimportant in the grooming rates of this community.

Table 4.7 Numbers of Grooming Interactions among Eight Adult Chimpanzees (*Pan troglodytes*) Studied in both 1976–1977 and 1982–1983

	TA	Ka	Nn	Fn	Jr	Vl	Yo	Pm
				1976–1977				
TA	—	2	3	2	1	0	1	0
Ka	1	—	3	1	6	1	3	0
Nn	1	3	—	0	3	1	1	16
Fn	3	2	0	—	6	0	1	3
Jr	1	1	1	5	—	0	1	14
Vl	0	0	1	0	0	—	1	1
Yo	1	3	3	0	1	0	—	0
Pm	1	1	0	2	7	0	0	—
				1982–1983				
TA	—	5	0	1	2	1	1	3
Ka	2	—	0	1	10	0	0	4
Nn	1	0	—	1	3	0	0	4
Fn	0	1	1	—	3	2	0	6
Jr	1	7	0	0	—	5	0	7
Vl	0	0	1	1	10	—	1	5
Yo	4	2	0	1	4	5	—	13
Pm	0	2	3	7	5	2	2	—

Rows indicate groomers and columns indicate groomees.
From Sugiyama (1988).

Table 4.8 Models Fit to Grooming Rates among Eight Adult Chimpanzees given in table 4.7

Factors included							
Identity of groomer	Identity of groomee	Dyad	Time period	Parameters estimated	AIC	ΔAIC	
X				9	266.0	35.7	No support
	X			9	230.3	0.0	**Best**
X	X			16	244.6	14.3	No support
		X		50	332.4	102.1	No support
X			X	10	268.1	37.7	No support
	X		X	10	232.0	1.7	Some support
X	X		X	17	246.3	15.9	No support
		X	X	51	334.3	103.9	No support

The best model, indicated by the lowest Akaike information criterion (AIC), is indicated in bold. The lowest AIC indicates the best-fitting model, and ΔAIC (difference between AIC and that of best model) indicates the degree of support for the other models.

This community-wide modeling and testing of interaction rate data seems to move away from the stated purpose of this section, examining dyadic relationships, but the models help because they produce expected interaction rates under the various hypotheses. These can be compared with real interaction rates, allowing us to decide whether the actual interaction rate of a dyad is unexpectedly high or low and thus to assess the relationship of its members.

4.8.2: Asymmetric and Symmetric Relationships. An *asymmetric relationship* can be defined as one in which the members of a dyad interact with one another at "significantly" different rates, so perhaps A grooms B more than B grooms A. I have placed quotes around "significantly" because the term can have two meanings: biological significance—important variation from the perspective of the individual—or statistical significance—variation in the observed measure beyond that expected by chance alone if there was really no difference. These are not the same. With large sample sizes, statistical significance can be found when there is no biological significance, and with small sample sizes, a biologically significant effect may not be detected statistically. In the context of asymmetry, I mean biological significance. Asymmetry in relationships can have large biological significance, for instance, skewing mating opportunities so that only some individuals within a society breed, and consequently reducing the effective genetic population size.

To measure asymmetry in a relationship, we need one or more asymmetric interaction measures (Section 3.2). A simple measure of asymmetry for one interaction measure is the difference in interaction rates divided by the sum of the interaction rates, as proposed by Beilharz and Cox (1967):

$$a_{AB} = \frac{I_{AB} - I_{BA}}{I_{AB} + I_{BA}} \tag{9}$$

where I_{AB} is the interaction rate, or number of interactions, between actor A and receiver B, and I_{BA} is that between actor B and receiver A. This measure of asymmetry varies between $a_{AB} = 0.0$, indicating equal rates in both directions and a symmetric relationship, and $a_{AB} = 1.0$ in which case A is always the actor and B always the receiver, or $a_{AB} = -1.0$, in which case B is always the actor and A always the receiver. van Hooff and Wensing's (1987) "directional consistency index" is simply the absolute value of a_{AB} (i.e., always positive).

An alternative in the case in which I_{AB} is a count of asymmetric interactions is de Vries et al.'s (2006) dyadic dominance index:

$$D_{AB} = \frac{I_{AB} + 0.5}{I_{AB} + I_{BA} + 1} \tag{10}$$

D_{AB} is a Bayesian estimator of the probability that individual A "wins" an encounter with individual B, and it potentially ranges from 0.0 to 1.0.

D_{AB}, a_{AB}, or some other index of asymmetry has little value without some measure of precision. If the interaction rates are calculated using independent sampling periods (Section 4.4), then bootstrap (resampling sampling periods with replacement) or jackknife (omitting each sampling period in turn) are suitable (Section 2.3). An analytical formula for the approximate SE of a_{AB} can be calculated using the delta method [a standard method in applied statistics (Tietjen 1986, p. 61)] on Equation (9):

$$\text{SE}(a_{AB}) = \frac{2\sqrt{I_{AB}^2 \, \text{Var}(I_{BA}) + I_{BA}^2 \, \text{Var}(I_{AB})}}{(I_{AB} + I_{BA})^2} \tag{11}$$

If I_{AB} and I_{BA} are counts of observed interactions that can be considered independent, then they should be Poisson distributed, and thus $\text{Var}(I_{AB}) \approx I_{AB}$, so that Equation (11) reduces to

$$\text{SE}(a_{AB}) = \frac{2\sqrt{I_{AB} \, I_{BA} \, (I_{AB} + I_{BA})}}{(I_{AB} + I_{BA})^2} \tag{12}$$

I have tested this formula on simulated data and it works well.

For the dyadic dominance index [Equation (10)], the delta-method estimate of the SE becomes

$$\text{SE}(D_{AB}) = \frac{\sqrt{(I_{AB} + 0.5)^2 \, \text{Var}(I_{BA}) + (I_{BA} + 0.5)^2 \, \text{Var}(I_{AB})}}{(I_{AB} + I_{BA} + 1)^2} \tag{13}$$

If the interactions can be assumed to be independent and Poisson distributed then, in Equation (13), $\text{Var}(I_{AB})$ can be replaced by I_{AB} and $\text{Var}(I_{BA})$ by I_{BA}.

If the interaction measures are calculated using independent subjects, then we can test for statistically significant asymmetry using likelihood-ratio G or chi-squared tests (Sokal & Rohlf 1994, 686–697), which are almost equivalent. Assuming, once again, that I_{AB} and I_{BA} are counts of observed interactions, then we can write

$$G = 2\left[I_{AB} \log_e(I_{AB}) + I_{BA} \log_e(I_{BA}) - (I_{AB} + I_{BA}) \log_e\left(\frac{I_{AB} + I_{BA}}{2}\right) \right] \tag{14}$$

$$X^2 = \frac{(I_{AB} - I_{BA})^2}{I_{AB} + I_{BA}}$$

Table 4.9 Frequency of Grooming (Vervaecke et al. 2000) in a Captive Group of Six Bonobos (*Pan paniscus*), with Calculated Measures of Asymmetry [Equation (9)], SEs [Equation (12)] in parentheses, and Results of Likelihood-Ratio G tests of Null Hypothesis That Grooming Is Symmetric [Equation (14)]

	Dz	He	Des	Ho	Lu	Ki
			Frequency of grooming			
Dz	0	24	41	11	1	0
He	22	0	1	19	0	0
Des	48	8	0	40	2	0
Ho	83	63	40	0	9	1
Lu	23	6	15	55	0	6
Ki	89	19	8	9	6	0
			Asymmetry a_{AB} (SE)			
Dz						
He	−0.04					
	(0.07)					
Des	0.08	0.78				
	(0.05)	(0.10)*				
Ho	0.77	0.54	0.00			
	(0.03)**	(0.05)**	(0.06)			
Lu	0.92	1.00	0.76	0.72		
	(0.04)**	(0.00)**	(0.08)**	(0.04)**		
Ki	1.00	1.00	1.00	0.80	0.00	
	(0.00)**	(0.00)**	(0.00)**	(0.09)**	(0.14)	

The groomers (actors) are the rows, and the groomees (receivers) are the columns.
* $P < 0.05$; ** $P < 0.01$.

These statistics (G or X^2) are compared to the χ^2 distribution with one degree of freedom, and the null hypothesis (no asymmetry) is rejected if they are unexpectedly high.

As an example of these methods, Table 4.9 shows the measure of asymmetry a_{AB} and its standard error for the 15 dyadic grooming relationships in a captive population of six bonobos. Also shown are the results of likelihood-ratio G tests for asymmetry. The grooming relationships in this population vary dramatically. There are pairs who rarely groom one another (e.g., Des and He), pairs who groom each other frequently and symmetrically (e.g., Des and Dz), and very asymmetric pairs in which grooming is almost entirely in one direction (e.g., Dz and Ki).

If several asymmetric interactions are measured, then a_{AB} could be calculated for each, and they could be combined in various ways. Because the sign of a_{AB} is in some respects arbitrary, depending on how the interaction is defined, then it may make sense to use the absolute values (ignoring minus signs) when combining them [i.e., the directional consistency index of van Hooff and Wensing (1987)], for instance, as a straight arithmetic mean of the a_{AB}'s on the different interaction measures.

4.8.3: Dominance Relationships. Dominance relationships are a major part of many, perhaps most, vertebrate social structures. Although interactions and associations define dominance, very often whether and how individuals interact and associate are strongly affected by their relative dominance. Drews (1993) defined *dominance* as "an attribute of the pattern of repeated, agonistic interactions between two individuals, characterized by a consistent outcome in favor of the same dyad member and a default yielding response of its opponent rather than escalation." Thus, dominance is a particular and strong form of asymmetry in a relationship. For a dyad, the options are that A is the consistent winner and so dominates B, B dominates A, or there is no dominance. Furthermore, we might wish to quantify the dominance.

Dyadic dominance can be measured in several ways (Lehner pp. 1998, 328–330). Experiments can be conducted in which a pair competes for a limited resource introduced by the experimenter and the individual that wins the resource is noted. This is also sometimes possible in a nonexperimental field (or captive) setting in which the resources are those naturally competed for by members of the population. Dominance can also be inferred by examining the asymmetry of aggressive and/or submissive behavior in dyadic encounters. If, when A and B are together, A wins contests for limited resources more than B does, or A displays more aggressive behavior and/or B more submissive behavior, then we can conclude that A is dominant over B.

In many cases, the determination is trivial: One individual wins the vast majority of contests and shows greatly more aggressive behavior and much less submissive behavior. However, dominance can be less complete or less obvious. In these cases, the techniques used to examine asymmetry in relationships, as discussed in the preceding subsection, are directly relevant and useful. A measure of asymmetry, such as a_{AB} proposed by Beilharz and Cox (1967) or, probably better, de Vries et al.'s (2006) dyadic dominance index [Equation (10)] indicates the degree of dominance. Using the results of statistical tests of asymmetry [such as the G test, Equation (14)] (Lehner 1998, p. 330) to assign dominance is conceptually invalid because it confuses biological significance with statistical significance.

Although dominance, if present, is a vital element of any dyadic relationship, most dominance relationships are sufficiently conspicuous that methodological development has mostly focused on the next step, delineating dominance hierarchies (considered in Section 5.4).

4.8.4: Dependent Relationships. Another type of asymmetric relationship is *dependency*, in which one member of a dyad depends on another for the necessities of life, usually food or protection from predators. Dependent relationships are usually between parent and offspring and are usually sufficiently obvious (did the young mammal suckle or the young bird beg?) that little collection or processing of data is required. Observations of interactions that indicate dependency, however, such as suckling or begging, may be used to trace changes in dependency, especially around the crucial transition from dependency (weaning or fledging).

4.8.5: Reciprocal Relationships. Behavioral ecologists consider that there are three principal processes leading to cooperation among animals: mutualism, kinship and reciprocity (Krebs & Davies 1991, p. 265; Section 7.3). Although mutualism, in which both individuals benefit, is not much of a puzzle, and kinship is based in the genes, *reciprocity* ("I'll scratch your back if you scratch mine") stems from social behavior itself (Trivers 1971). Thus, we may seek reciprocity in data on dyadic social interactions. It is both impracticable and unrealistic, however, to examine reciprocity in a relationship without reference to other relationships. One could, theoretically, look at the temporal arrangement of asymmetric interactions in one relationship to examine reciprocity (e.g., "Is A more likely to assist B soon after B has helped A?"), but I know of no such investigations. Partly this is because the general model of reciprocity is time averaged and partly because the conceptual models of reciprocity include relativity with respect to other relationships. Hemelrijk (1990b) considered two dyadic models of relative reciprocity that she found in the behavioral literature:

1. *Actor-reactor model.* Individuals give relatively more to those individuals that give to them relatively more in return compared to what they give to other individuals: I am Joe's best friend, so he can be mine.

2. *Actor-receiver model.* Individuals give more to those individuals from whom they receive more: I receive most from Joe, so I will help him most.

These models seem very similar but they differ importantly in both their practical tractability and what they assume about animals' mental processes. Hemelrijk (1990b) shows that the actor-receiver is both easier to fit to real data and makes fewer assumptions about the cognitive abilities of the animals. It assumes that animals can keep track of what they receive from whom, whereas the actor-reactor model assumes knowledge of the nature of all other relationships in the community. Thus, in what follows, I assume an actor-receiver model of reciprocity as the alternative to the null hypothesis that the frequency of dyadic asymmetric interactions with A the actor and B the receiver has no direct dependence on the frequency when B is the actor and A is the receiver.

Under this and other models, reciprocity is relative, and so we cannot look at the reciprocity of one relationship in isolation. We can, however, look at relationships involving one individual. In Table 4.9, the bonobo Ho grooms the other members of its community at rates of 83, 63, 40, 9, and 1 per unit time and receives grooming from the same individuals at rates of 11, 19, 40, 55, and 9, respectively. A test of actor-receiver reciprocity is that these two series are correlated. The correlation coefficient is $r = -0.37$, however, indicating no positive relationship between the rates at which Ho grooms others and the rate she is groomed by them, and so there is no evidence for actor-receiver reciprocity in grooming relationships for Ho. We can modify the hypothesis so that it only considers ranks of grooming or being groomed by others, such that Ho grooms more often those individuals from whom she receives more grooming. In this case, a test statistic is Spearman's rank correlation coefficient r_s. For the grooming relationships of Ho, $r_s = 0.0$, so again there is no sign of reciprocity.

This approach can be extended to the community level, using variants of the Mantel test (Hemelrijk 1990b; see Section 2.4). The basic idea is to compare the actor-receiver interaction matrix with its transpose, the receiver-actor matrix. To transpose a matrix, one flips it about its diagonal so that the rows become columns and the columns become rows (Table 4.10). If there is reciprocity, then the high elements of the actor-receiver matrix should correspond to the high elements of the receiver-actor matrix, so that when the interaction level within a dyad is high in one direction, then it is also high in the other. The relationship between the actor-receiver interaction matrix and its transpose can be tested using the Mantel test (Hemelrijk 1990b; Section 2.4), with statistically significant positive values of the matrix correlation coefficient indicating reciprocity. As Hemelrijk (1990b) explains, however, properly to test the actor-receiver model of reciprocity, we need to use a variant of the Mantel test, the R_r test, in which the actor-receiver matrix and its transpose, the receiver-actor matrix, are first ranked within rows (Table 4.10). Using the bonobo example (Table 4.9), we see that the row in the actor-receiver matrix that represents Ho grooming others (83, 63, 40, 9, 1) is transformed into ranks 1, 2, 3, 4, 5, and that in the

Table 4.10 Actor-Receiver and Receiver-Actor (Transpose) Matrices for Bonobo Grooming Data (table 4.9), Together with Ranked-within-Rows Matrices

Actor-receiver matrix						Receiver-actor matrix (transpose)					
0	24	41	11	1	0	0	22	48	83	23	89
22	0	1	19	0	0	24	0	8	63	6	19
48	8	0	40	2	0	41	1	0	40	15	8
83	63	40	0	9	1	11	19	40	0	55	9
23	6	15	55	0	6	1	0	2	9	0	6
89	19	8	9	6	0	0	0	0	1	6	0
↓						↓					
Ranked						Ranked					
0	2	1	3	4	5	0	5	3	2	4	1
1	0	3	2	4.5	4.5	2	0	4	1	5	3
1	3	0	2	4	5	1	5	0	2	3	4
1	2	3	0	4	5	4	3	2	0	1	5
2	4.5	3	1	0	4.5	4	5	3	1	0	2
1	2	4	3	5	0	4	4	4	2	1	0

The matrix correlation ($r = 0.12$), and Mantel test ($P = 0.033$) between the two lower matrices constitute Hemelrijk's (1990a) R_r test for reciprocity.

receiver-actor matrix that represents Ho being groomed by others, 11, 19, 40, 55, 9 becomes 4, 3, 2, 1, 5 (Table 4.10). After we rank these matrices within rows, the R_r test follows the Mantel test. For the bonobo data, the matrix correlation of the R_r test is 0.12 with significance $P = 0.33$, so there is little support for reciprocity in this community.

Hemelrijk (1990b) discusses a number of variants of this approach to testing for reciprocity, including the use of an analog of Kendall's rank correlation coefficient, the K_r test, instead of the R_r test, and testing for reciprocity between two different classes of individual (e.g., do males reciprocate grooming by females?). She also considers testing for absolute reciprocity and qualitative reciprocity. In absolute reciprocity, all individuals value acts identically, and to test for this, Hemelrijk (1990b) suggests ranking the entire off-diagonal actor-receiver and receiver-actor similarity matrices (rather than ranking within rows as with the R_r test), a procedure known as the R test (Dietz 1983). In qualitative reciprocity, we are only concerned with whether an interaction did or did not occur, not its frequency, and so all nonzero interaction rates are replaced by one in actor-receiver and receiver-actor matrices before the Mantel test.

Another form of reciprocity occurs when the exchange is between different types of act: "I'll scratch your back if you bring me food." Thus, the actor-receiver and receiver-actor matrices are for different types of interaction and are not transposes of one another. Hemelrijk (1990b) calls this "interchange" and shows how the R_r and R tests can be used in these cases. A problem, particularly with interchange analyses, is that

the relationship between the two acts might not be causal but due to
one or more other attributes that relate to the two measured behavioral
rates, such as dominance rank. In a second paper, Hemelrijk (1990a)
shows how to use partial correlation methods to tease out such effects,
for instance, testing for a relationship between grooming and support
in agonistic interactions, controlling for differences in dominance rank.

Additional possibilities, considered in Section 7.3, are that reciprocity is not dyadic. Individual A does not adjust its behavior toward
B based on B's behavior toward A, but based either on B's behavior
toward the population in general, "reputation reciprocity" ("I am good
to those who are good to others") (Mohtashemi & Mui 2003), or the
general population's behavior to A, "generalized reciprocity ("If I have
received, I will give") (Pfeiffer et al. 2005).

4.9 *"Special" Relationships: Permutation Tests for Preferred/Avoided Companionships*

Are certain relationships especially strong or weak? There are numer-
ous examples in the literature on animal social structures of overly interpreted results: Large association indices are considered bonds, small
ones indicate a pair that avoids one another, and so on. Even with random association within a population, there will be one dyad that has
the largest association index and one that has the smallest. Before interpreting relationship measures as a bond, avoidance, or anything similar
(Section 4.10), it is highly desirable to compare them with those from
randomly interacting or associating individuals (Bejder et al. 1998), in
other words, to use the data to test the null hypothesis that, given certain constraints, individuals associate at random against alternatives
that individuals have preferred or avoided companions.

The constraints usually considered concern the data structure. The
number of individuals identified during each sampling period, the number of identifications of each individual, or similar characteristics of the
data are taken as fixed. At this level of analysis, we are interested in
neither the pattern of field effort nor differences in identifiability between animals, and we wish to factor them out of any conclusion about
preferred or avoided companionships. We may also wish to control for
the gregariousness (Section 4.3) of the individuals.

The problem can then be set up as a permutation test (Section 2.4).
We consider a statistic that indicates the degree of variation in dyadic
association within a community, such as the standard deviation of asso-

ciation indices (Section 4.5), and calculate it for the real data. We then make many random permutations of the data, subject to constraints in the pattern of identifications of individuals (such as those suggested in the previous paragraph), calculating the statistic for each. If the real statistic is greater than 95% of those from the random data, then we may conclude that there are indications of preferred and/or avoided companions within the community at $P < 0.05$. This test is one sided because it is hard to envisage a scenario that would produce less varied association indices than expected by chance.

The challenge is with the permutation. It is not trivial to permute association data randomly subject to constraints such as keeping constant both the number of identifications in each sampling period and the number of identifications of each individual. Early attempts (e.g., Smolker et al. 1992; Whitehead et al. 1982), although successful, used cumbersome computational techniques that would not necessarily work on other data sets. A major breakthrough occurred when Bejder et al. (1998) harnessed a technique that Manly (1995) had developed for the congruent ecological problem of determining whether pairs of species cooccur more frequently than would be expected. Manly noted that whereas producing new independent random data sets subject to the constraints is challenging, new, nonindependent data sets can be constructed easily, and if this is done enough times, the distribution of test statistics is equivalent to that from independent random permutations. The procedure is illustrated using a group-by-individual matrix in Table 4.11. The constraints are keeping constant both the number of individuals in each group and the number of groups in which each individual was observed. This is achieved by choosing two individuals and two groups randomly so that each individual is identified in just one of the groups and each group contains just one of the individuals. Then the four group-individual assignments are flipped (in Table 4.11, the individual A in group r is now in group k and vice versa for individual F), preserving the totals for each group and individual. Thus, we have a new data set with the constraints preserved. This is only slightly different from its precursor, and they are not independent of one another. The new data set in turn can be flipped into another data set, and so on. Although permutations produced with flips are not independent because a flip makes only a small change to the data matrix (Table 4.11), Manly (1995) showed that this does not matter as long as sufficient flips are carried out. Typically, such tests use many more than the 1,000 permutations or so that are usual for permutation tests.

Table 4.11 One Flip in Bejder et al.'s (1998) Process of Permuting a Group–Individual Matrix

Group	A	B	C	D	E	F	G		Group	A	B	C	D	E	F	G
a	1	1	0	0	0	0	0		a	1	1	0	0	0	0	0
b	1	1	1	1	0	0	0		b	1	1	1	1	0	0	0
c	1	1	0	0	0	0	0		c	1	1	0	0	0	0	0
d	1	1	1	1	0	0	0		d	1	1	1	1	0	0	0
e	0	0	1	0	1	1	1		e	0	0	1	0	1	1	1
f	0	0	1	0	0	1	1		f	0	0	1	0	0	1	1
g	0	0	0	0	1	1	1		g	0	0	0	0	1	1	1
h	0	0	0	0	0	1	1		h	0	0	0	0	0	1	1
i	0	1	0	0	0	0	1	⇒	i	0	1	0	0	0	0	1
j	0	1	0	0	0	1	1		j	0	1	0	0	0	0	1
k	0	1	0	0	0	1	0		k	1	1	0	0	0	0	0
l	1	0	1	0	0	0	0		l	1	0	1	0	0	0	0
m	0	0	1	0	1	0	0		m	0	0	1	0	1	0	0
n	1	0	0	0	1	0	0		n	1	0	0	0	1	0	0
o	1	0	0	0	1	0	0		o	1	0	0	0	1	0	0
p	0	0	1	1	0	0	0		p	0	0	1	1	0	0	0
q	1	0	0	1	0	0	0		q	1	0	0	1	0	0	0
r	1	0	1	1	0	0	0		r	0	0	1	1	0	1	0
s	1	0	1	1	1	0	0		s	1	0	1	1	1	0	0

The matrix on the left, showing which groups contained which individuals, is modified by randomly choosing two individuals and two groups (with each individual in only one of the groups and each group containing only one of the individuals) and switching assignments (shaded), preserving row and column totals. Table adapted from Whitehead (1999a).

An attractive feature of Bejder et al.'s (1998) procedure is that one can investigate the null hypothesis of no preferred or avoided companionship for any dyad as well as for the entire community. If the real association index of a dyad is greater than 95% of the random indices for that dyad, then we may conclude that the dyad is a preferred companionship at $P < 0.05$. Unlike the overall test for preferred/avoided companions in the community, which is one sided, it often makes sense to use two-sided tests for dyadic association, so we test simultaneously for preferred companionship (a significantly high value of the association index) or avoided companionship (a significantly low value of the association index).

Dyadic P values must be used carefully. It typically takes many more permutations to stabilize dyadic P values than the P value for the test against random companionship for the entire community (Whitehead et al. 2005). In addition, with N individuals in the community, there will be $N(N − 1)/2$ dyadic P values and so a large number of false-positive test results if the P values for all dyads are examined in even a moderate-sized community. This can be addressed using Bonferroni or other multiple-comparisons procedures (Bejder et al. 1998), but then power is considerably reduced. I do not think any dyadic P values should be considered if the overall, community-wide test for preferred/avoided companionships is not statistically significant or nearly statistically significant, or if fewer than 5% of the dyadic tests are significant (assuming a nominal significance level of 0.05). A final warning with regard to dyadic P values is that they should not be used as measures of the strength of a relationship and input into further analyses (Whitehead et al. 2005). Doing so confuses statistical significance with effect size. A dyadic P value depends on the strength of the relationship but also on sample sizes and data structure (the data for some dyads may not be amenable to much permutation, giving the test little power). Methods of quantifying the strength of a relationship are considered in Section 4.10.

There are several variants on the Bejder et al. (1998) method. These allow testing for preferred/avoided companionship in particular but important situations (Whitehead 1999a; Whitehead et al. 2005). In the next subsections, I consider these different variants, the circumstances in which they can be used, the assumptions and practical issues involved, such as the choice of a test statistic, and, finally, some notes on technical issues. Table 4.12 summarizes the Bejder et al. (1998) test, its variants, and this advice.

These procedures have uses beyond the testing of hypotheses about preferred or avoided companionship. They produce "random" data sets and "expected" association indices that can be most useful when constructing models of social structure (Sections 4.10, 5.3, and 5.7).

4.9.1: Permuting a Group-by-Individual Matrix. In the original form of the test proposed by Bejder et al. (1998), the data can be represented by a group-by-individual matrix, as in Table 2.2, and these are permuted as indicated in Table 4.11. A variety of test statistics is possible. Bejder et al. (1998), following Manly (1995), suggested using the sum of squares of the differences between the observed and expected half-weight association indices, in which the expected association index between any pair is the mean over all the random data sets. Alternative test statistics with the advantage that they have a more intuitive meaning in terms of social structure are the standard deviation of the association indices and the standard deviation, mean, and median of the nonzero association indices. After trying a number of these and other statistics, I found that it usually matters little which is chosen and that the coefficient of variation of association indices is perhaps most intuitive and least affected by data structure (Whitehead et al. 2005; Table 4.12).

Table 4.12 Summary of Permutation Tests for Preferred/Avoided Companionships: Bejder et al. (1998) Test and Variants

	Permuting:		
	Group-by-individual matrix	Group-by-individual matrix within sampling periods	Matrices of association indices
Used for	Groups defined; closed population	Groups defined; open population	Associations defined; open or closed population
Suggested test statistics, short term	—	Prop(AI⁻); mean(AI)	—
Suggested test statistics, long term	CV(AI); SD(AI); SD(AI⁻); mean(AI⁻); median(AI⁻)	CV(AI); SD(AI); SD(AI⁻)	SD(AI)
Controls for movement into and out of the study area	No	Yes	Yes
Controls for differences in gregariousness	No	No	Yes
Important considerations	Independence of groups Closed population Number of flips	Independence of groups Length of sampling periods Number of flips	Independence of sampling periods Length of sampling periods Number of flips

AI, association index; AI⁻, nonzero association indices; Prop, proportion.

Important assumptions of this permutation method are that groups are independent and that the population is closed. If groups are not independent, so that records of groups close together in time are more likely to contain the same individuals than groups further apart, then the permutation test is more likely to reject the null hypothesis of no preferred/avoided companions than the nominal level of the test if there is no effect.

A cause for rejection of the null hypothesis could be differences in gregariousness among individuals. Suppose that some "asocial" individuals are found only in small groups, whereas other, "hypersocial" individuals generally seek out large groups. The former will tend to have low association indices and the latter high ones, so that these differences will increase the test statistic, indicating variability in association indices above that expected if all individuals had the same affinity for groups of a particular size. The group-by-individual permutations remove any characteristic individual gregariousness, lowering the expected test statistic. Thus, differences in gregariousness can lead to rejection of the null hypothesis of no preferred/avoided companions, even though the only preference or avoidance is indirect: Because of their preference for large groups, hypersocial individuals will be more likely than expected to be associated with each other and less likely than expected to be associated with asocial individuals.

If the population is not closed, so that some individuals are not present during part of the study period because of birth, death, or emigration, then they will not be available for sampling during these periods. Dyads with similar patterns of presence in the study area will tend to have high association indices, and those who were rarely in the study area together will have low association indices, even if there is no preferred/avoided companionship. Thus, once again, statistically significant results of the permutation test may occur even when there is no preference or avoidance, in this case because of demographic effects.

4.9.2: Permuting a Group-by-Individual Matrix within Sampling Periods. A way around the problem of demographic effects is to make the flips of the group-by-individual matrix only within sampling periods within which there is likely to have been little birth, death, immigration, or emigration (Whitehead 1999a; Table 4.13). Thus, the analysis is constrained for the number of times each individual was identified in each sampling period rather than the number of times it was identified overall.

With this method, a suitable length of sampling period must be chosen. It should be short enough so that animals rarely join or leave the population within sampling periods, but long enough so that there are several groups within most sampling periods, allowing identifications to be flipped in a number of ways. If sampling periods contain few data, then few flips are possible and the test loses power.

In this situation, we can test for preferred/avoided companions both between (long-term) and within (short-term) sampling periods. Simulations suggested that the proportion of nonzero association indices and the mean of all association indices are suitable test statistics when testing for preference/avoidance within sampling periods, with low real values of the statistics indicating preferred or avoided companions within periods (Whitehead et al. 2005). If there are preferred/avoided companionships within sampling periods in the real data, then there will be proportionally more pairs of individuals that are repeatedly grouped within sampling periods than expected. Thus, because the number of individuals in each group is constrained, proportionally fewer dyads are grouped during sampling periods and overall, decreasing the proportion of nonzero association indices and the mean association index. To test for preferred/avoided companionships between sampling periods,

Table 4.13 One Flip in the Variant of Bejder et al.'s (1998) Process of Permuting a Group-Individual Matrix in Which Demographic Effects Are Controlled

Group	Individuals A	B	C	D	E	F	G		Group	Individuals A	B	C	D	E	F	G
a	1	1	0	0	0	0	0		a	1	1	0	0	0	0	0
b	1	1	1	1	0	0	0		b	1	1	1	1	0	0	0
c	1	1	0	0	0	0	0		c	1	1	0	0	0	0	0
d	1	1	1	1	0	0	0		d	1	1	1	1	0	0	0
e	0	0	1	0	1	1	1		e	0	0	1	0	1	1	1
f	0	0	1	0	0	1	1		f	0	0	1	0	0	1	1
g	0	0	0	0	1	1	1		g	0	0	0	0	1	1	1
h	0	0	0	0	0	1	1		h	0	0	0	0	0	1	1
i	0	1	0	0	0	0	1	⇒	i	0	1	0	0	0	0	1
j	0	1	0	0	0	1	1		j	0	1	0	0	0	1	1
k	0	1	0	0	0	1	0		k	0	1	0	0	0	1	0
l	1	0	1	0	0	0	0		l	1	0	1	0	0	0	0
m	0	0	1	0	1	0	0		m	0	0	1	0	1	0	0
n	1	0	0	0	1	0	0		n	1	0	0	0	1	0	0
o	1	0	0	0	1	0	0		o	1	0	0	0	1	0	0
p	0	0	1	1	0	0	0		p	0	0	1	1	0	0	0
q	1	0	0	1	0	0	0		q	0	0	0	1	1	0	0
r	1	0	1	1	0	0	0		r	1	0	1	1	0	0	0
s	1	0	1	1	1	0	0		s	1	0	1	1	1	0	0
t	0	0	1	0	1	0	0		t	1	0	1	0	0	0	0

The data were collected in five sampling periods, divided by horizontal dashed lines. At this step, the fifth sampling period was randomly chosen, and within it, the matrix was modified by randomly choosing two individuals and two groups and switching assignments (shaded), preserving row and column totals within the sampling periods.
Method and table from Whitehead (1999a).

the coefficient of variation of the association indices seems a suitable test statistic, with high values indicating rejection of the null hypothesis of random association (Whitehead et al. 2005). The standard deviation of the association indices and the standard deviation of the nonzero association indices are also potential statistics for long-term, between-period preference/avoidance, but if there is a short-term effect, this will lower the mean association index and so its standard error, lessening the ability to detect long-term effects.

This version of the test, like the original, does not control for differences in individual gregariousness. The null hypothesis may be rejected in cases in which individuals have distinctive preferences for groups of particular sizes, even though they do not possess preferences for associating with particular individuals.

Table 4.14 One Flip in the Process of Permuting a Symmetric 1:0 Similarity Matrix for a Randomly Chosen Sampling Period

	Individuals									Individuals							
	A	B	C	D	E	F	G	H		A	B	C	D	E	F	G	H
A	–	1	0	0	0	1	0	1	A	–	1	0	1	0	1	0	0
B	1	–	1	1	0	0	1	1	B	1	–	1	1	0	0	1	1
C	0	1	–	1	1	1	1	0	C	0	1	–	0	1	1	1	1
D	0	1	1	–	0	1	0	0	D	1	1	0	–	0	1	0	0
E	0	0	1	0	–	0	1	1	E	0	0	1	0	–	0	1	1
F	1	0	1	1	0	–	0	0	F	1	0	1	1	0	–	0	0
G	0	1	1	0	1	0	–	1	G	0	1	1	0	1	0	–	1
H	1	1	0	0	1	0	1	–	H	0	1	1	0	1	0	1	–

(⇒ between the two matrices at row D)

The symmetric 1:0 matrix on the left is modified by randomly choosing two pairs of individuals (D and H; A and C) so that each individual is associated with only one member of the other pair. The associations between these pairs (shaded) are switched, preserving row (and column) totals.
Method and table from Whitehead (1999a).

4.9.3: Permuting Matrices of Association Indices. Bejder et al.'s (1998) original test and the variant that deals with demography by flipping within sampling periods assume that association is defined using groups, but this need not be the case; association can be defined using criteria such as spatial proximity or behavioral interactions, which do not lead to transitive groups (Section 3.3). These data can be represented by a series of matrices of associations, one for each sampling period (as in Table 2.3). To test for preferred/avoided companionship in such situations, flips are made within 1:0 symmetric matrices of associations for randomly chosen sampling periods, with two pairs of individuals being chosen such that each member of one pair is associated with just one member of the other pair. The associations are then switched (Table 4.14).

A suitable test statistic when permuting matrices of association indices is the standard deviation of the association indices. This tests for preferred/avoided companionship between sampling periods; there is no test for short-term, within-sampling-period association in this case. Sampling period length must be chosen carefully. Associations should be independent between sampling periods, and sampling periods should be sufficiently long so that they contain enough data so that a range of possible flips can be made. Very short sampling periods will produce matrices of association indices mostly containing zeros (few associations), and long sampling periods may produce matrices mostly containing ones (most pairs associated). In either case, there are few possible flips and little power to the test.

This procedure has the advantage over its predecessors in that it does control for differences in individual gregariousness. In each sampling period, the number of associates of each individual is held constant. This feature, as well as its applicability to any definition of association and control over demographic effects, makes this version of the test, in which the associations themselves are permuted, especially attractive. Its drawbacks are that it generally needs rather more data than the group-by-individual permutations and it only addresses long-term, between-sampling-period association preferences (Table 4.12).

4.9.4: Some Technical Issues with the Bejder et al. and Related Tests. In addition to the choice of association index (Section 4.5), sampling period length, and test statistic (covered previously), there are several other technical issues that need to be considered when carrying out the Bejder et al. (1998) test or its variants. Because the flips are not independent, many more are needed to provide stable P values than in normal permutation tests. How many more? Unfortunately, there is no guide to how many flips are needed. Instead, it is recommended that one start with perhaps 10,000, note the P value(s), try 20,000, see whether the P value(s) are similar, and, if they are not, continue adding more flips until they do stabilize. Then, once P values seem to have stabilized at a certain number of flips, repeat with that number of flips to check their stability. Note that many more permutations (flips) are usually required to stabilize dyadic P values than the overall P value for the community (Whitehead et al. 2005).

Each flip only changes the data, and so the test statistic, a little (Table 4.11). Hence, it is partially redundant to calculate the test statistic after every flip as proposed by Manly (1995). We have found it most efficient computationally to calculate test statistics after about every 100 flips (Whitehead et al. 2005), and this is what SOCPROG allows one to do.

All of these tests can also be carried out between classes of individual, so we may ask whether there are differences among the relationships between individual males and individual females. These are done by simple restrictions to the Bejder et al. (1998) test or its variants (Whitehead et al. 2005).

4.10 *Quantifying the Strength of a Relationship and the Bond*

Reiterating a point made at the start of the previous section, dyadic significance tests should not be used to measure the strength of a relationship. Instead, the association index, as a measure of the proportion of time a dyad spends in association (Section 4.5), is an indicator of the strength of a relationship, whereas the P value of the dyadic permutation test (Section 4.9) is an indicator of confidence that the association is unusually large or small (Table 4.15).

Examples of this approach are shown in the matrices of association indices for disk-winged bats (Table 4.16) (Vonhof et al. 2004) and mature male northern bottlenose whales (Table 4.17) (Gowans et al. 2001). In the former case, the associations are often very obvious. Some pairs of bats always roosted together and some never did so. The null hypothesis of random association was clearly rejected by the permutation test for the whole matrix and for each of the high-association dyads. These bats seem to have strong friends, and the analysis identifies them.

The results of the analysis for the mature male bottlenose whales are less clear cut (Table 4.17). Although the overall test for preferred/avoided associations is statistically significant and there are some quite large and quite small association indices, only three are statistically significant at $P <$ 0.05, and the actual coefficient of variation of the association indices (CV = 1.64) is only a little larger than the expected value (CV = 1.52). The bottlenose whales seem to have preferred companions, but the analysis does not shed much light on how strong they are or which they are.

Sometimes, we may wish to divide relationships into those that are strong (perhaps "friends") and those that are weak ("acquaintances"), perhaps for input to a binary network analysis (Section 5.3). As I have emphasized, the dyadic P value is not a good basis for doing this (and see Table 4.17). Any other measure will be somewhat arbitrary. One basis my colleagues and I have used (Durrell et al. 2004; Gero et al. 2005) is to define "friends" as those dyads with association indices greater than twice the average index, so that these are pairs of individuals that are associated at least twice as much as the expected value of a dyad chosen randomly from the community (Table 4.15).

Association indices themselves, or dichotomous measures based on them such as greater or less than twice the mean association index, do not consider gregariousness (Section 4.3). Thus, a very gregarious animal is likely to have many friends, even though there may not be any mutual attraction between them (Pepper et al. 1999). Pepper et al. (1999) suggest using the ratio of the observed to the expected dyadic association index as a measure of friendship corrected for gregariousness, in which the expected dyadic association index might be calculated using the Bejder et al. (1998) permutation method or one of its variants (Section 4.9). These "ratio indices," which correct for gregariousness, could then be

Table 4.15 Measures That Indicate the Strength of a Relationship

Measure	What is it?	Affected by	Dichotomous version	Recommendation
Association index	Estimate of proportion of time associated	Sampling method, demography, gregariousness	Greater than twice the mean community-wide association index	Straightforward; index should be chosen to reduce biases from sampling methods and demography
Observed association index÷expected association index	Proportionally, how much more or less the dyad is associated than expected, given its gregariousness		> 2.0	Good; compensates for gregariousness, sampling methods, and demography (if appropriate method is used to calculate expected values); complex to calculate
Significance of association index from permutation test	Probability of obtaining a more extreme association index than for random association, given sampling scheme, gregariousness, and demography (if appropriate method is used)	Sample size, data structure	$P < 0.05$	Not recommended as a measure of the strength of a relationship

Dichotomous versions (e.g., "friends" and "acquaintances") are suggested.

Table 4.16 Association Indices among 10 Disk-Winged Bats (*Thyroptera tricolor*) Based on at Least 6 Days of Identification

	al102	al115	al138	al140	al23	al81	al86	al87	al88	y22
al102										
al115	0.00									
al138	0.00	0.00								
al140	0.00	0.00	**1.00****							
al23	0.00	0.00	**1.00****	1.00						
al81	0.00	0.00	0.00	0.00	0.00					
al86	0.00	0.00	**0.57****	**0.57****	0.57	0.00				
al87	0.00	0.00	**0.71****	**0.71****	**0.71****	0.00	**0.83****			
al88	**1.00****	0.00	0.00	0.00	0.00	0.00	0.00	0.00		
y22	0.00	0.00	0.00	0.00	0.00	0.00	0.00	0.00	0.00	

See Vonhof et al. (2004). The mean simple ratio association index, using sampling periods of days and groups defined by roosts, is 0.19. Dyads with association indices at least twice the mean are shown in bold. Double asterisks indicate dyadic $P < 0.01$ for the original Bejder et al. (1998) test with 1,000 permutations and 100 flips per permutation, for preferred association (no dyadic P values were $0.01 < P < 0.05$, and there were no significantly low association indices at $P < 0.05$). The overall significance for preferred/avoided associations in this matrix, using the coefficient of variation (CV) of association indices as a test statistic, is $P = 0.001$ (CV = 1.82 for real data, compared with CV = 1.67 for mean of random permutations).

used to produce dichotomous measures. For instance, "friends" could be defined as dyads who spend twice at least as much time associated as would be expected, given their gregariousness (Table 4.15).

The association index of a dyad is affected by factors other than their "friendship" and the gregariousness of the two individuals. These include their identifiabilities, the sampling methods, and demographic effects (who could be associated with whom, given patterns of presence in the study area?). Methods of reducing biases due to these effects are discussed in Section 4.5. Pepper et al's (1999) ratio of observed to expected index is particularly useful because if a suitable permutation method is used to calculate the expected indices, this can go a long way toward correcting for sampling and demographic effects as well as for gregariousness (Table 4.15). For instance, the Bejder et al. (1998) permutation method corrects for sampling effort by permuting groups, or permutes associations within sampling periods so that demographic effects are corrected for, whereas if associations are permuted, then gregariousness is also controlled (Table 4.12).

4.10.1: Bonds. *Bonds* are frequently considered crucial elements of social structures (e.g., Wrangham 1980) and social behavior (e.g., Zahavi 1977), yet the term "bond" is rarely defined. It seems to connote more than mutual attraction, and there is usually an implied element of temporal stability: Bonded animals interact preferentially over long time periods. Usually, however, even more than this is implied.

Table 4.17 Association Indices among 16 Mature Male Bottlenose Whales (*Hyperoodon ampullatus*) Based on at Least 7 Days of Identification

	#1	#3	#10	#16	#33	#37	#71	#118	#120	#225	#413	#480	#804	#824	#950	#1114
#1																
#3	**0.31***															
#10	**0.09**	0.00														
#16	0.00	0.00	0.00													
#33	**0.11**	0.00	**0.35**	0.05												
#37	0.07	0.03	0.03	**0.12**	**0.11**											
#71	0.00	0.00	0.00	0.00	0.00	0.03										
#118	0.03	0.00	0.00	0.00	0.00	0.00	0.00									
#120	0.05	0.10	0.00	0.00	0.00	0.00	0.00	0.00								
#225	0.07	0.00	0.05	0.00	0.00	0.07	0.07	0.00	0.00							
#413	**0.13**	**0.08**	**0.24**	0.00	**0.25**	**0.10**	0.00	0.00	0.00	0.00						
#480	0.07	**0.08***	0.05	**0.08**	**0.10**	0.00	0.00	0.00	0.00	0.00	0.00					
#804	0.07	0.03	0.04	0.00	**0.20**	**0.15**	0.00	0.05	0.04	0.05	**0.09**	0.04				
#824	**0.11**	**0.11**	**0.08**	0.00	0.07	0.05	0.00	**0.18***	0.03	0.04	**0.08**	**0.13**	0.03			
#950	0.00	0.00	0.00	0.00	0.00	0.00	0.00	0.00	0.00	0.00	0.00	0.00	0.04	0.00		
#1114	0.00	0.00	0.00	0.00	0.00	0.00	0.00	0.00	0.00	0.00	0.00	0.00	0.00	0.00	0.00	

See Gowans et al. (2001). The mean simple ratio association index, using sampling periods of days and association defined by identification within 15 min, is 0.039. Dyads with association indices at least twice the mean are shown in bold. Asterisk indicates dyadic $P < 0.05$ for test for preferred association using permutation of associations within sampling periods (10,000 permutations with 100 flips per permutation). The overall significance for preferred/avoided associations in this matrix, using the coefficient of variation (CV) of association indices as a test statistic, is $P = 0.001$ (CV = 1.64 for real data, compared with CV = 1.52 for mean of random permutations)

Smuts (1985) noticed that male–female dyads of olive baboons (*Papio cynocephalus*) that had relatively high grooming rates also were likely to be found close together. Thus, for any female, she defined "affiliates" as those males who were ranked first or second for both grooming rates and proximity scores. Following this, R. Wrangham (personal communication; see also Wrangham et al. 1992), suggests that bonded animals are those that have strong relationships on two or more "independent" interaction or association measures. Here, I do not mean statistical independence, which would be indicated by an effectively zero correlation between the dyadic values of the association or interaction measures; the presence of bonded dyads, which have high values on both of the measures, would likely lead to a positive correlation. Instead, the measures should be behaviorally independent. Thus, the rates at which animals make courtship displays and the rates at which they copulate would not be independent, whereas the rates at which they make courtship displays and the rates at which they groom might be considered independent.

An example, for female capuchin monkeys, is shown in Fig. 4.5. Among the six females in the community, the two dyads with the highest grooming scores were also those that most often formed coalitions against other females. Thus, by Wrangham's criterion, we might consider these two dyads to have bonds.

I have suggested the use of multiple strong association or interaction measures as a way objectively and quantitatively to identify bonds, but am not sure that it is the ideal approach and am fairly certain it will not be in some situations. I look forward to developments.

4.11 Relationships between Classes

As Hinde (1976) notes, considerable insight into social structures comes from abstractions of relationships among pairs of individuals to classes of individual (Fig. 1.4). We can generalize from the rates of interaction, or association indices, between pairs of females to get a mean and standard deviation for female–female relationships and can compare these with male–female relationships, and so on. Sometimes, the classes used in these abstractions arise from the social analysis itself (Section 3.6). For instance, if permanent, or fairly permanent, social units are identified (Section 5.7), then the interactions and associations among social units can be studied.

Calculating interaction rates within and between classes might seem straightforward: How many interactions were observed between males

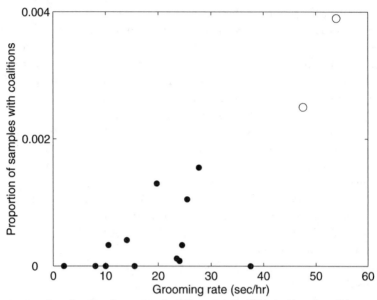

FIGURE 4.5 Grooming rate and proportion of sampling periods in which a dyad formed a coalition against other females for dyads among six female capuchin monkeys (*Cebus capucinus*). The two dyads with high rates of both grooming and coalition formation (unfilled circles) might be considered bonded. (Redrawn from Perry 1996.)

and males, between females and females, and between males and fe-males? If there are different numbers of individuals in each class, how-ever, which there usually will be, and the effort spent observing indi-viduals varies, which it often will, then corrections are needed, as will cases in which the population composition changes during the course of the study. Altmann and Altmann (1977) discuss how to make these corrections. Table 4.18 gives an example showing how rates of ago-nistic interaction change with time and between sex classes for ground squirrels.

Interclass associations are examined in two principal ways. We can ask: What are the mean number of members of class B associated with a member of class A at any time, or what is the mean association index between members of class A and B? The latter is easily calculated from dyadic association indices. The former, which I call interclass gregar-iousness, is an extension of the concept of typical group size (Section 4.3; also see later discussion). It may also be useful to consider the mean proportion of associates belonging to a particular class, which is the ra-tio of interclass or intraclass gregariousness to typical group size minus one (Underwood 1981).

Table 4.18 Interaction Rates among Richardson's Ground Squirrels (*Spermophilus richardsonii*) by Time Period in 1975 and Sex Class

Two-week time period		Interaction rates per overlapping pair per survey		
		F-F	F-M	M-M
1	Late pregnancy	0.025	0.009	0.052
2	Lactation	0.024	0.014	0.021
3	Lactation	0.036	0.028	0.026
4	Postweaning	0.023	0.011	0.016
5	Postweaning	0.022	0.002	0
6	Postweaning	0.005	0	0
7	Prehibernation	0	0	—
8	Prehibernation	0	—	—

Rates are only calculated for dyads with overlapping home ranges.
From Michener (1980).

Table 4.19 Mean Simple Ratio Association Indices (SD) for Dyads among Age-Sex Classes of 63 Bottlenose Whales (*Hyperoodon ampullatus*) Identified on at Least 5 Days

	Female	Subadult male	Mature male
Female	0.011 (0.006)		
Subadult male	0.007 (0.011)	0.047 (0.059)	
Mature male	0.008 (0.010)	0.013 (0.012)	0.037 (0.024)

Simple ratio association index, using sampling periods of days and association defined by identification within 15 min. A Mantel test of the null hypothesis that intraclass and interclass association indices had the same mean was rejected ($p < 0.0001$).

Whereas interclass gregariousness and intraclass gregariousness generally differ (because of different numbers of individuals in each class), realistic null hypotheses are that interaction rates or association indices do not differ between and within classes. Such hypotheses can be tested using the Mantel test (Section 2.4). A matrix correlation between the matrix of dyadic interaction rates, or association indices (e.g., Table 2.5), and a 1:0 matrix indicating whether members of a dyad are from the same or different class is calculated. The classes of the individuals are randomly permuted and the matrix correlation recalculated for each random permutation, allowing the statistical significance of the real matrix correlation to be assessed (Section 2.4; Mantel 1967; Schnell et al. 1985).

Table 4.19 gives an example of using association indices to examine interclass associations for bottlenose whale sex classes. Associations within sex classes are substantially higher, especially for males.

Most of the other methods described in this chapter for analyzing dyadic relationships can be abstracted to classes. For instance, interclass

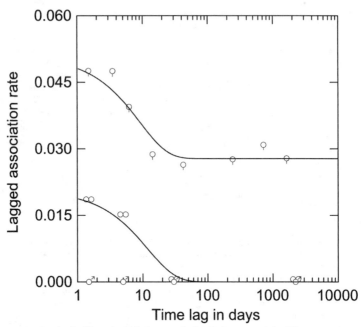

FIGURE 4.6 Standardized lagged association rates for Ecuadorian sperm whales (*Physeter macrocephalus*) over time periods from 1 day to several years for pairs of mature males (♂), females (♀), and female–male dyads (oo). For the male–female and female–female data, curves are fitted based on exponential decay in the probability that individuals continue associating (Section 5.5). (Redrawn from Whitehead 1997.)

asymmetry, dominance, and reciprocity can be calculated from the means of the appropriate dyadic values (Section 4.8). We can also produce lagged interaction/association rates (Section 4.6) within and between classes. Figure 4.6 plots standardized lagged association rates within and between sex classes of sperm whale. The classes have very different types of relationship: Females show permanent (within the several-year scale of the study) associations with one another as well as more temporary acquaintances lasting about 1 week. Male–female associations last about 1 week and then break down, whereas no individual male was observed to associate with another individual male over more than 1 day.

It may also be instructive to plot inter- and intraclass relationship measures in multidimensional space. Gero et al. (2005) calculated association indices between bottlenose dolphins separately for different behavioral states. Figure 4.7 plots associations when foraging or feeding against those when socializing; these relationships are indexed by age/sex class. Gero et al. (2005) distinguished four general types of relationship: affiliates who were associated strongly (i.e., dyads who spent most of their time together in both behavioral states), acquaintances

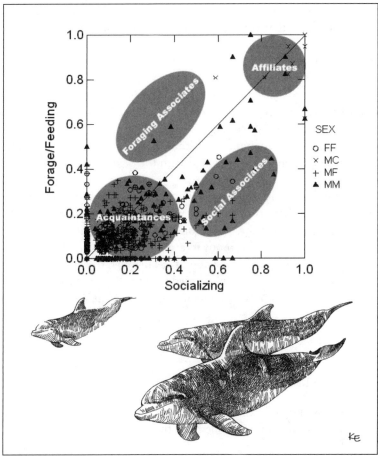

FIGURE 4.7 Relationships between dyads of bottlenose dolphins (*Tursiops* spp.) in Shark Bay, Western Australia (data from Gero et al. 2005), classified by sex, and indicating four general types of relationship, as indicated by association index when socializing (*x* axis) and foraging (*y* axis). FF, between females; MC, mother-calf; MF, male-female; MM, between males. (Illustration copyright Emese Kazár.)

who were weakly associated in both states, social associates who spent much more time together when socializing, and the unusual relationship of foraging associates who spent much more time together when foraging. Mother–calf dyads were affiliates, as were some relationships between two adult males. The only two foraging associate relationships, and most social associates, were between adult males. Dyads including females were mostly of the acquaintance type. Plotting the distribution of interclass and intraclass associations as in Fig 4.7 gives a much clearer insight into the distribution of relationships than simply presenting mean interclass and intraclass values.

In long-term studies, analytical power and insight may be achieved through analyses that compare the relationships of identified individuals before and after they change class. Such transitions could include weaning or sexual maturity.

4.11.1: Animals Assigned to Class but Not Individually Identified. It is possible to make some inferences about relationships between classes in situations in which individuals are not individually identified but can be assigned to classes. Interaction rates between classes can be calculated, although some care is needed with the denominator (Altmann & Altmann 1977).

If group composition is recorded whenever groups are encountered, then, assuming that groups are equally likely to be encountered whatever their composition, interclass and intraclass gregariousness and the mean proportion of companions in a particular class can be calculated using the equations of Underwood (1981). Thus, if $n_X(k)$ is the number of members of class X in observed group k, the mean number of companions of class Y with a member of class X (XY gregariousness) is

$$G_{XY} = \frac{\sum\limits_{k} n_X(k) \cdot n_Y(k)}{\sum\limits_{k} n_X(k)}$$

For intraclass gregariousness (the mean number of companions of the same class), we just subtract one from the intraclass typical group size:

$$G_{XX} = \frac{\sum\limits_{k} n_X(k)^2}{\sum\limits_{k} n_X(k)} - 1$$

Overall gregariousness for a class (the mean total number of companions) is calculated similarly:

$$G_X = \frac{\sum\limits_{k} n_X(k) \cdot \left(\sum\limits_{Y} n_Y(k) - 1 \right)}{\sum\limits_{k} n_X(k)} = \sum\limits_{Y} G_{XY}$$

The estimated proportion of companions of members of class X that are of class Y is then G_{XY}/G_X. Pepper et al.'s (1999) "pairwise affinity" index, which completely controls for the gregariousness of both classes, is

$$\frac{\left[\sum_k n_X(k) \cdot n_Y(k)\right] \cdot \left[\sum_k \sum_Z n_Z(k) \cdot \left(\sum_Z n_Z(k) - 1\right)\right]}{\left[\sum_k n_X(k) \cdot \left(\sum_Z n_Z(k) - 1\right)\right] \cdot \left[\sum_k n_Y(k) \cdot \left(\sum_Z n_Z(k) - 1\right)\right]}$$

Similarly, the intraclass pairwise affinity index is

$$\frac{\left[\sum_k n_X(k) \cdot (n_X(k) - 1)\right] \cdot \left[\sum_k \sum_Z n_Z(k) \cdot \left(\sum_Z n_Z(k) - 1\right)\right]}{\left[\sum_k n_X(k) \cdot \left(\sum_Z n_Z(k) - 1\right)\right]^2}$$

As an example of some of these measures, for Underwood's (1981) eland (*Taurotragus oryx oryx*), in January, the prime mating season, mean interclass and intraclass gregariousness between classes were estimated as follows:

> Adult males with average adult female: 2.5
> Subadult males with average adult female: 0.7
> Adult males with average adult male: 1.5
> Subadult males with average adult male: 0.5
> Adult females with average adult male: 4.6

Gregariousness among other pairs of classes is not given. The proportions of companions of an adult male are 23% other adult males, 8% subadult males, and 70% adult females; and the overall gregariousness is 6.6 companions for each adult male.

Finally, if we know, or have estimates of, the number of individuals in each class in the community N_X, then we can determine the mean interclass and intraclass association indices (estimates of the proportion of time spent together) from $\alpha_{XY} = G_{XY}/N_Y$. Thus, because there were about 60 female eland in the population studied by Underwood, an estimate of the mean association index between individual adult males and individual females is $4.6/60 = 0.077$.

These are examples of the rather few useful measures of social structure that can be calculated without identifying individuals.

BOX 4.2 *Describing Relationships: Recommendations*

This chapter summarizes a wide range of techniques for describing relationships. There seem to be many options. For some species, however, collecting interaction or association data is sufficiently challenging that we can go little beyond group-based association indices (Section 4.5) as a relationship measure. Even here, however, there are choices, including the length of sampling period (Section 3.9) and type of association index (Section 4.5). With more easily viewed species, the options increase dramatically. My overall advice in choosing relationship measures has two components:

1. Think from the individual animal's perspective. What appear to be the important time scales of social interactions and relationships, and in what ways are an individual's social relationships manifested? The heuristic answers to these questions can be used to select appropriate relationship measures.

2. Produce as many relationship measures as both seem appropriate for the animals and are technically feasible without compromising precision. It does not matter whether they are correlated or even redundant so that one relationship measure contains basically the same information as another. The fact that they are correlated is interesting in its own right, and such redundancy can be removed from further analyses using multivariate techniques (Section 5.6).

5

Describing and Modeling Social Structure

The third stage of Hinde's (1976) conceptual framework for the study of social structure is to use the content, quality, and patterning of relationships to describe social structure. The description is sometimes qualitative, for example, "Core social units of related females are accompanied by unrelated males." The addition of quantitative measures, however, adds much, for example, "Core social units of 2 to 6 females with a mean relatedness of 0.13 are accompanied by 1 to 2 males, who are unrelated either to each other or the females they are accompanying. Each male stays with a unit for a mean of 14 months." Visual displays of social structure usually enhance and inform such descriptions. Qualitative, quantitative, and visual techniques of analyzing social structure can be purely descriptive, letting the data speak for themselves, or may use one or more models. We can often use the collected data to estimate parameters of the model that might be behaviorally informative (such as mean group size). It is also helpful to assess the validity of the model using goodness-of-fit and other measures, as well as to compare fits of different models.

The most appropriate method of describing a social structure depends heavily on its attributes and on the data available. For instance, social structures involving many thousands of animals cannot be represented usefully by dendrograms, but these are often highly appropriate for smaller populations. In this chapter, I suggest appropriate methods for displaying and modeling social structures of different types, whether highly structured or loose and labile, and whether containing 6 individuals or 6,000.

The chapter begins by considering simple quantitative (Section 5.1) and visual (Section 5.2) descriptors of social structure that use one-dimensional relationship measures. More complex, and usually informative, techniques often use models, and I introduce network analysis (Section 5.3), dominance hierarchies (Section 5.4), temporal analysis using lagged association rates (Section 5.5), and multivariate analyses of two or more relationship measures (Section 5.6). The final section considers the assignment of individuals to communities, social units, or other social entities (Section 5.7). All these methods are principally within the ethological, descriptive perspective. Chapter 6 and especially Chapter 7 focus more directly on the "why?" questions of a behavioral ecologist.

BOX 5.1 *Omitting Individuals from Analyses of Social Structure*

In many studies, especially field studies, information on the webs of relationships among individual animals—their social niches— varies considerably. Some individuals, perhaps those whose ranges are closest to the base of the field workers or who have lived through a large part of a long-term study, will have been observed many times and will often have been recorded interacting or associating. The data used to describe their relationships are good, and will form a strong basis for describing or modeling the social structure of the population. In contrast, other individuals, perhaps more peripheral in range or short lived, may have been rarely observed. Knowledge of their relationships will be sketchy. When describing and modeling social structures, if we include these poor-data individuals, perhaps it will bias our perspective or model of the society. This is an important issue.

For a few of the methods described in this chapter, the data from poorly studied individuals have no disproportionate effect on the outcome of the analysis: lagged association rates (Section 5.5) and Bayesian and temporal methods of unit delineation (Section 5.7). In the former case, the estimated lagged association rate is an integration of the available data, with that from poorly studied animals having appropriately small impact. The Bayesian method will indicate great uncertainty in the unit delineation of poorly studied animals, and, usually, much less uncertainty for well-studied animals, whereas temporal unit delineation will not assign individuals to units if there are insufficient data.

All the other methods considered in this chapter are based on matrices of one or more relationship measures: interaction rates or association indices. For some of the dominance indices or ranking methods, individuals with relatively little data available are given less weight, but they are given an index or rank (Section 5.4). However, for other methods—univariate attributes of social structure (Section 5.1), ordinations (Section 5.2), cluster analyses (Section 5.2), network analyses (Section 5.3), and multivariate methods (Section 5.6)—the interaction rates and association indices go into the analysis without any indication of their accuracy, thus potentially allowing poorly studied individuals to unduly influence the output, perhaps giving a misleading or biased model of social structure.

So, often, the social analyst should consider omitting some poorly studied individuals from an analysis (while perhaps retaining them for other analyses, such as lagged association rates). But which to omit? If the data are divided into sampling periods (Section 3.9), then the number of sampling periods observed makes a suitable measure of the data available for an individual. So we might make ordinations for individuals identified in at least 5, 10, or 20 sampling periods. But which cutoff should we use?

The precision of interaction rates [Equation (5)] and that of association indices [Equation (6)] both decrease inversely as the square root of the number of observations. They also depend, however, on the rates of association or interaction: With rare associations or interactions, considerable data are needed for their pattern to emerge. Thus, I cannot give a recommendation such as "For analysis Y, just use individuals identified more than x times."

One can say, however, that interaction rates or association indices based on four or fewer samples will always be inaccurate. Therefore, a minimum cutoff of using only individuals identified in five sampling periods for any analysis that does not account for variation in data quality seems sensible. I suggest also trying sequentially larger cutoffs of perhaps 10 and 20 sampling periods to see whether results change substantially as the lower-quality data are excluded.

5.1 Attributes of Social Structure

In this section, I list and discuss a number of measurable or estimable attributes that help the social analyst describe the social structure of a population. Many follow from Wilson's (1975) 10 "qualities of sociality," which are reproduced and annotated in Section 1.8 (see also Table 1.2). This list is distinct from the "attributes of the data set" of Section 3.10, although the nature of the data set determines which attributes of social structure can be measured or estimated and how well.

> *Modularity.* A study population may be socially homogeneous or well divided. Modularity (described in Section 5.7) indicates how well a population can be delineated into communities or social units. Newman (2004) suggests that a modularity greater than about 0.3 (on a scale in which 0.0 indicates random association and 1.0 indicates no associations between closed units or communities) indicates important divisions. The modularity of a population may vary, however, with the definition of association (Section 3.3), the choice of sampling period (Section 3.9), the type of association index (Section 4.5), and how expected associations are calculated (Section 5.7).

> *Community size.* As used in this book, a community is a set of individuals that is largely behaviorally self-contained and within which most individuals interact with most others. From an individual's perspective, community size is important because it approximates the number of other individuals that it may interact with (Section 1.8) and so might need to distinguish (Dunbar 1998). A study population may contain one or more communities, or the study may focus on just part of a community. In some situations, such as the loose societies of pelagic fish or migratory ungulates or territorial species in which individuals principally interact with their neighbors, there may be no community. Methods of deciding whether a population contains one or more communities and of allocating individuals to communities are described in Section 5.7, and techniques of estimating community size when not all individuals are identified are summarized in Appendix 9.5. If there are communities, the following attributes should usually be calculated within each.

Rates of interaction and association (typical group size). We
can estimate the mean rates, over dyads and individuals,
of interaction per time unit (Section 4.4) or association
per sampling period (mean association index; Section 4.5)
for all available measures. If association is defined using
groups, then the mean over individuals of the sum of association indices translates into gregariousness or typical
group sizes minus one (Section 4.3). High interaction rates
indicate active societies, and high association rates indicate large numbers of animals associated or grouped at any
time.

Social differentiation. A useful attribute of the social structure
of a community is its social differentiation—how much variation there is in dyadic probability of association. I suggest
that this variation be expressed by the coefficient of variation
of the true association indices, in other words the standard
deviation, over dyads, of the true proportion of time spent
associated, divided by the mean proportion of time spent associated. Appendix 9.4 gives two methods of estimating this
from real data [Equation (23) gives the maximum likelihood
formula; Equation (24) gives a simpler but less accurate and
more biased estimate]. The precision of this measure is probably best assessed using the bootstrap procedure in which
sampling periods are resampled with replacement (Section
2.3). There are a number of reasons why true association
indices may vary in addition to preferred/avoided companionships, and these will increase social differentiation. These
include demographic effects (death, birth, and migration
causing some pairs to be present together more than others)
and associations within the community being structured by
factors that could include dominance, dependency, kinship,
range use, age or gender class, gregariousness, or permanent
social units.

Asymmetry. If asymmetric interaction measures are available,
then the mean value of some measure of asymmetry [such
as van Hooff and Wensing's (1987) directional consistency
index, the absolute value of a_{AB}; Section 4.8] can be calculated.

Linearity of dominance. Measures of dyadic dominance such
as interaction rates that give the winners of fights, access to
competed resources, or dominant/submissive behavior can

be used to calculate Landau's (1951) index of dominance linearity h or de Vries' (1995) h' (Section 5.4). High values indicate communities with a linear structure, such as a dominance hierarchy, and low values indicate a more egalitarian society (Section 5.4).

Stability of associations. If association measures are available, then the temporal stability of associations can be indicated by measures such as the mean half-life of associations, which can be estimated from the time lag at which the lagged association rate equals 0.5 (Whitehead 1997; Section 5.5). This is a sensible approach with fairly simple societies, but when things become more complex, such as brief associations among permanent units, then more complex measures or displays are needed to capture the temporal nature of relationships.

Network measures. Means and standard deviations of some network analysis measures of dyads or individuals (Section 5.3), such as strength, reach, clustering coefficient, and affinity, can be informative, as can measures calculated for an entire community, such as diameter.

Class structure. All of the aforementioned attributes can be separated into intraclass and interclass measures, such as the numbers of individuals of each class in the community, interaction and association rates within and between classes, and social differentiation within and between classes.

Table 5.1 illustrates the use of some of these attributes for northern bottlenose whales off Nova Scotia.

5.2 Single-Measure Displays of Social Structure

A relationship measure, such as an interaction rate or an association index, is dyadic. Potentially, it gives a value to each dyad in the population. These can be listed as a matrix of numbers, such as those in Tables 2.5, 4.4, 4.16, and 4.17. The human eye and brain are not particularly good at assimilating large arrays of numbers, so that the essence of the social structures displayed in these matrices is obscure, even with just seven individuals as in Table 2.5, and if the table is very large, say with 100 individuals, and so 4,950 elements (assuming a symmetric measure and matrix; 9,900 if not), the pattern is virtually incomprehensible as it stands. Thus, we seek to display the matrix in a graphical form that

Table 5.1 Attributes of the Social Structure of a Population of Northern Bottlenose Whales (*Hyperoodon ampullatus*) off Nova Scotia

Attribute	Population	Females/immatures (F)	Mature males (MM)	Subadult males (SM)
Population size (Whitehead & Wimmer 2005)	163	79	46	38
Mean association index	0.012	0.009 [F–F 0.011 F–MM 0.008 F–SM 0.007]	0.016 [MM–F 0.008 MM–MM 0.037 MM–SM 0.013]	0.014 [SM–F 0.007 SM–MM 0.013 SM–SM 0.047]
Social differentiation: coefficient of variance of true association indices	0.74	0.70	0.94	0.95
Modularity-G (number of communities/units detected; section 5.7)	0.448 (10)	0.512 (7)	0.241 (4)	0.106 (2)
Network measures (section 5.3):				
Strength	0.67	0.55	0.95	0.51
Reach	0.61	0.43	0.97	0.44
Clustering coefficient	0.11	0.10	0.13	0.12
Affinity	0.82	0.72	0.97	0.89

Social attributes are calculated from 63 individuals as in table 4.19

is more appropriate for human cognitive abilities. In this section, I discuss methods of displaying matrices of association indices or interaction measures. I illustrate the values and drawbacks of each method by trying to use them on four populations with very different attributes:

1. Observations of seven black-capped chickadees at a feeder. The data (Ficken et al. 1981) are of half-weight association indices indicating presence at a feeder together. Like other matrices of association indices, the data matrix is symmetric.

2. Rates of grooming among six female capuchin monkeys [shown in Table 4.4, from Perry (1996)]. This is an asymmetric interaction rate matrix.

3. Observations of 55 disk-winged bats roosting together in leaves [part of the data are shown in Table 4.16 (Vonhof et al. 2004)], using simple ratio association indices and restricting consideration to individuals identified on 5 or more days.

4. Associations among 353 sperm whales identified off the Galápagos Islands between 1985 and 2002, with association defined as observed within 1 hour of each other, sampling periods of days (only individuals identified on 3 or more days are included), and a half-weight index (Whitehead 2003).

The social structure of each of the four populations was displayed (when possible) using a histogram of association indices or interaction rates, a sociogram, a principal coordinates analysis, nonmetric multidimensional scaling and a hierarchical cluster analysis (Figs. 5.1–5.4). I discuss each of these methods using the displays of the four populations as a guide, concluding the section with a mention of other methods and types of display.

5.2.1: Histograms of Association Indices or Interaction Rates. A simple display of a matrix of association indices or interaction rates is a histogram showing their distribution. This is informative with almost any relationship measure or study population. Overlaying histograms for relationships between different pairs of classes (as in Figs 5.1, 5.3, and 5.4) usually helps. For instance, associations among chickadees are generally strongest between sexes, but this is not the case for the bats, who seem to have no preference to associate with the same or a different sex, or in sperm whales, in which all the strong associations are among female–

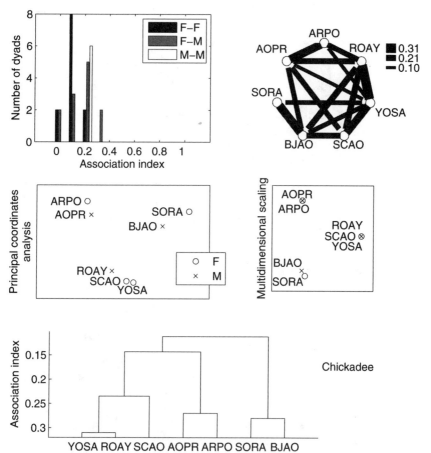

FIGURE 5.1 Ordinations of half-weight association indices of observations of chickadees (*Parus atri-capillus*) at a feeder. Distribution of association indices, sociogram, principal coordinates analysis (first two coordinates explaining 42.8% of the variance in the matrix of association indices), nonmetric multidimensional scaling (squared stress criterion; stress = 0.0018), and average-linkage cluster analysis (cophenetic correlation coefficient = 0.77). (Data from Ficken et al. 1981.).

female dyads. If there are many individuals in the population (as in Figs 5.3 and 5.4), it is often clearer to log the y-axis because otherwise the all-important high-association dyads are swamped by the many dyads with near-zero associations. Although histograms do summarize the distribution of relationships within a study population, they do not describe their structure. Quite different social structures can produce similar-looking histograms. For instance, there might be a bimodal distribution

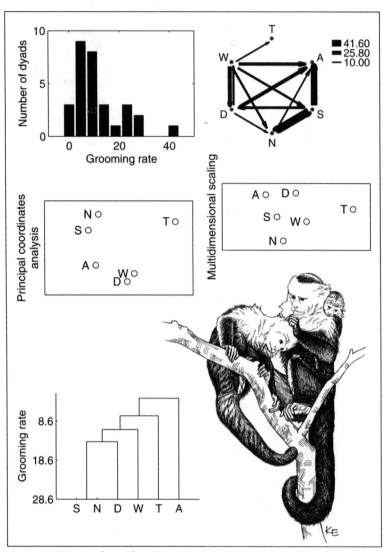

FIGURE 5.2 Ordinations of rates of grooming among six female capuchin monkeys (*Cebus capucinus*) (table 4.4). Distribution of grooming rates, sociogram (arrow points from groomer to groomee), principal coordinates analysis (first two coordinates explaining 73.0% of the variance in the matrix of grooming rates), nonmetric multidimensional scaling (squared stress criterion; stress = 0.00048), and average-linkage cluster analysis (cophenetic correlation coefficient = 0.94). For the last three displays, the association between any pair of monkeys is the average of their unidirectional grooming rates. (Data from Perry 1996.) (Illustration copyright Emese Kazár.)

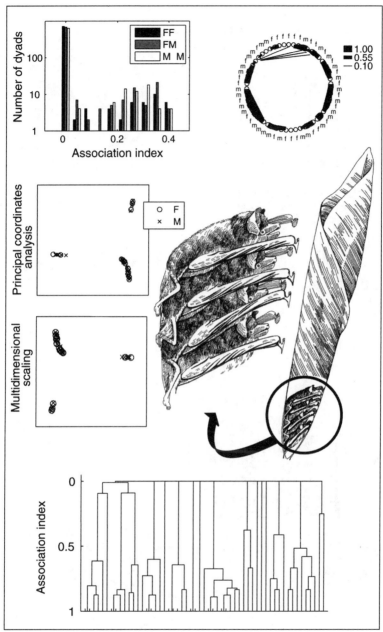

FIGURE 5.3 Ordinations of simple-ratio association indices of observations of 55 disk-winged bats (*Thyroptera tricolor*) roosting together in leaves (Vonhof et al. 2004), just using individuals identified on 5 or more days. Distribution of association indices, sociogram, principal coordinates analysis (first two coordinates explaining 21.7 percent of the variance in the matrix of association indices), nonmetric multidimensional scaling (squared stress criterion; stress = 0.0018), and average-linkage cluster analysis (cophenetic correlation coefficient = 0.99). (Illustration copyright Emese Kazár.)

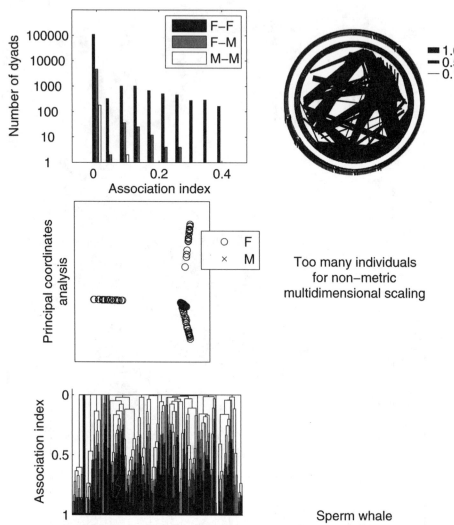

FIGURE 5.4 Ordinations of simple-ratio association indices of 353 sperm whales (*Physeter macrocephalus*) identified off the Galápagos Islands between 1985 and 2002 (Whitehead 2003), with association defined as observed within 1 hour of each other, sampling periods of days (only individuals identified on 3 or more days are included), and a half-weight index. Distribution of association indices, sociogram, principal coordinates analysis (first two coordinates explaining 7.7 percent of the variance in the matrix of associations), and average-linkage cluster analysis (cophenetic correlation coefficient = 0.92).

of association indices when permanent social units within which individuals are strongly associated interact casually with other units. A similar bimodal pattern (a few high indices and many low ones) might be produced by a range-based society in which animals whose core ranges overlap associate strongly, whereas more spatially separated individuals

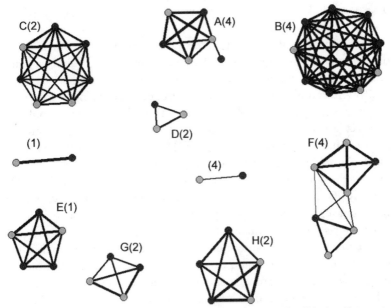

FIGURE 5.5 Sociogram of simple-ratio association indices of observations of 55 disk-winged bats roosting together in leaves (Vonhof et al. 2004), just using individuals identified on 5 or more days, drawn using NetDraw (black nodes are males; gray nodes are females), omitting bats unconnected by association indices less than 0.1. Clusters of bats are indicated by letters, and their locations among the three subunits of the study area (1, 2, 4) are indicated in parentheses.

have low association rates. To resolve such differences, we need other displays.

In some contexts (Section 6.2) it may be useful to rescale the y-axis of the histogram so that it indicates the mean number of associates of an individual rather than the number of associations in the population. This is done in Fig. 6.1.

5.2.2: Sociogram. The *sociogram* is a simple display of a relationship measure in which individuals are arranged as points (or nodes) in a plane, and lines showing links (or edges) are drawn between them indicating the strength of the relationship. In mathematical terminology, a sociogram is a "valued graph." In its implementation in SOCPROG, the individuals are equally spaced around the circumference of a circle, and the thickness of the linking lines indicates the size of the association index or interaction rate. It helps to place strongly related individuals close together (and SOCPROG tries to do this). Pairs of individuals with relationship measures less than some cutoff are usually not linked. If an asymmetric measure (often an interaction rate) is being examined, two directional links can be drawn between each dyad (e.g., Fig. 5.2). It is now a "directed valued graph."

Sociograms display almost all the data in matrices of association indices or interaction rates. For small populations (less than about 25 individuals), sociograms are an excellent, model-free way to display the data clearly and completely. They have been frequently used in the analysis of vertebrate social systems (Whitehead & Dufault 1999). As population sizes get larger, however, they become cluttered (e.g., Fig. 5.3) and then an overwhelming mess (e.g., Fig. 5.4). The usefulness of sociograms for displaying larger populations can be improved by thoughtful arrangement of the points representing the individuals. The circular style used by SOCPROG (and shown in Figs. 5.1–5.4) works well for small populations, but separating clusters of points representing highly associating or interacting sets of individuals—basically communities—or other designs can be much better for larger ones. Nonmetric multidimensional scaling (see later in this section) and network analysis programs (Section 5.3) may provide useful arrangements of the points. For instance, a more informative sociogram of the bat data of Fig. 5.3 is shown in Fig. 5.5 (using the same cutoff for indicating links, 0.10). It was drawn using NetDraw (Section 2.9) with initial positions of points from multidimensional scaling, which were then moved interactively to improve clarity. The bats form tight mixed-sex clusters, each well connected internally but poorly linked.

Network analysis (Section 5.3) goes on mathematically to analyze the properties of a sociogram.

5.2.3: Principal Coordinates Analysis. Principal coordinates analysis (sometimes called classic multidimensional scaling), a type of ordination, is described briefly in Section 2.6 (see also Digby & Kempton 1987, pp. 83–93; Manly 1994, pp. 190–198; Legendre & Legendre 1998, 424–444). The principal output is a display of points (sometimes called a "scores plot"), each representing an individual, in which the distances between the points is inversely related to their association index or interaction rate.

For association indices, the usually assumed ideal relationship is that the distance between two points is proportional to the square root of one minus the association index between the two individuals. Therefore, if two individuals have an association index of 0.0, their points should be plotted, say, 1 unit apart. If the index is 0.01, then they should be 0.9 units apart; if the index is 0.25, they should be 0.5 units apart; if it is 0.81, then they should be 0.1 units apart; and if the index is 1.0, they should be plotted on top of one another. With enough dimensions, such an arrangement of the points may be possible—or it may not. Suppose that the association index between A and B is 0.01, that between A and C is 0.16, and that between B and C is 0.81. Then the distances between the points are as follows: A–B 0.9, A–C 0.6, B–C 0.1. No such arrangement of points is possible in any number of real dimensions; the "triangle inequality" is violated. This is indicated mathematically by negative eigenvalues of the principal coordinates analysis. A few small negative eigenvalues and a few violations of the triangle inequality are probably not too serious, but having many large, negative eigenvalues indicates that principal coordinates analysis is not producing a useful ordination (Manly 1994, p. 194). There are ways to try to remove negative eigenvalues (Legendre & Legendre 1998, pp. 432–438). In such cases, however, I suggest trying nonmetric multidimensional scaling (see later discussion).

Principal coordinates analysis operates on symmetric matrices—there is only one distance between each pair of points, so only one value can be represented for each dyad. With asymmetric relationship measures, such as interaction rates, we can simply take the mean of the two associations, although potentially useful information is obviously lost. There is a related technique, the principal components biplot, which can be used for asymmetric matrices (Digby & Kempton 1987, pp. 151–154). Here, each individual is plotted twice, once as interactant and once as interactee. Interpretation of such plots, however, is not particularly straightforward, and I have never seen them used for social analysis.

With interaction rates, which, unlike association indices, are unbounded by an upper 1.0, the square root of one minus is not a suitable transformation as it stands. Instead, we can divide all the rates by the maximum rate before transforming using the square root of one minus.

Principal coordinates works mathematically by calculating the eigenvectors and eigenvalues of the dissimilarity or transformed similarity matrix. The eigenvectors—there are as many of them, n, as individuals—give the positions of the points representing individuals in n-dimensional space (Digby & Kempton 1987, pp. 83–84). The dimensions are arranged, as in principal components analysis (Section 2.6), so that as much as possible of the variance is explained by the distribution of the points in the first dimension, as much as possible of the remaining variance by the distribution of the points in the second dimension, and so on. This means that if we are interested in, for example, a three-dimensional representation, we take the first three eigenvectors or dimensions. The actual values of the points in the new coordinates (the axes of the plot) have no direct meaning and are usually of little interest. It is the arrangement of the points that should be informative.

With each eigenvector comes an eigenvalue, proportional to the amount of variance explained by the arrangement of the points in that dimension, and so the effectiveness of the principal coordinates display in any number of dimensions can be assessed. If perhaps 40% to 80% of the variance is explained by a principal coordinates analysis in a given number of dimensions, then some of the attributes of the social structure are being captured, but some are also being missed. With greater than about 80% of the variance explained, the display might be considered a very useful representation of the matrix of association indices.

Unfortunately, this is not often the case with social data, unless there are only a few individuals (Figs. 5.1 and 5.2) or a very large number of dimensions are being considered, and in this case the display is unwieldy. However, it is often worth trying a principal coordinates analysis. If it works, explaining a large proportion of the variance in a few dimensions with no large, negative eigenvalues, the analyst has a clear, uniquely determined display of the relationships in the community as described by the matrix of association indices. With large numbers (> ~50) of individuals, principal coordinates analyses are possible computationally (unlike multidimensional scaling) *and* can produce an informative display, in contrast to sociograms and cluster analyses.

An example of a fairly informative three-dimensional principal coordinates plot with a moderate number (27) of individuals is shown in Fig. 5.6 (with the three dimensions explaining 43% of the variance). The bats (a subset of those used in Figs. 5.3 and 5.5) appear to form five quite distinct co-roosting clusters. Some of the bats are plotted on top of one another, limiting the value of this display.

If principal coordinates analysis does not produce a useful display, for instance, because the first three dimensions explain less than perhaps about 40% of the variance, or there are large, negative eigenvalues, the next step may be to try multidimensional scaling, particularly the non-metric version. Multidimensional scaling has a similar output to principal coordinates analysis but produces it by a different method in which some of the conditions are relaxed.

5.2.4: Metric and Nonmetric Multidimensional Scaling. Morgan et al. (1976) advocate the use of multidimensional scaling for the analysis of non-human societies. Multidimensional scaling provides a display similar

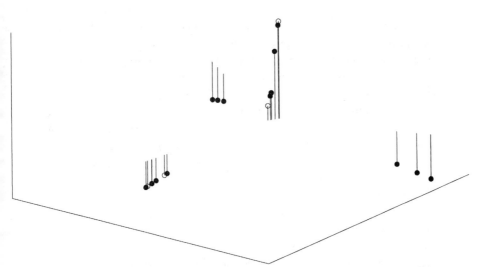

FIGURE 5.6 Three-dimensional ordination of simple-ratio association indices of observations of 27 disk-winged bats from two "locations" roosting together in leaves (Vonhof et al. 2004), just using individuals identified on 5 or more days, using principal coordinates analysis (first three coordinates explaining 43.3% of the variance in the matrix of association indices). Each circle representing an individual is linked to the $z = 0$ plane by a "spike." Males are represented by black circles, females by open circles.

to that of principal coordinates: a set of points in n-dimensional space arranged so that more associated dyads are plotted closer. Rather than employing the elegant but rather inflexible properties of eigenvectors to produce a plot, however, multidimensional scaling uses computational power. The analyst specifies, a priori, the number of dimensions required. A starting configuration in these dimensions is also given. This could be from principal coordinates or a random scattering of the individuals. Multidimensional scaling iteratively tries to find a "better" ordination by moving the points around. One of the useful attributes of multidimensional scaling is that "better" can be defined in a range of ways, giving different properties to the ordination. The "metric stress" criterion is similar to principal coordinates (Manly 1994, p. 171): The aim is to produce a display so that the distance between points is proportional to the transformed (usually using a "square root of one minus" transformation) association index. However, there are differences: Principal coordinates produces an optimal arrangement in all possible numbers of dimensions in one step; metric multidimensional scaling aims at a given number of dimensions and works iteratively, and so the optimal arrangement may not be achieved. This means that metric multidimensional scaling has few perceived advantages over principal coordinates analysis, and it is not often used.

In nonmetric multidimensional scaling (Legendre & Legendre 1998, pp. 444–450), the relationship between the matrix of association indices and distances between the points is relaxed to a monotonic one: More closely associated pairs of individuals are plotted more closely together than are less closely associated ones. "Nonmetric stress" represents the degree of failure in this criterion of a particular representation (Manly 1994, pp. 170–171; Legendre & Legendre 1998, p. 447), and nonmetric multidimensional scaling minimizes it. This is a much less onerous criterion than linear proportionality (used in principal coordinates analysis and metric multidimensional scaling), and so nonmetric multidimensional scaling can usually produce an acceptable ordination in fewer dimensions than principal coordinates or metric scaling. This is usually accomplished by separating the highly associated individuals more than in principal coordinates analysis.

Generally, nonmetric multidimensional scaling highlights the differences among the most associated individuals, and principal coordinates that among the least associated. This may be seen as an advantage of nonmetric multidimensional scaling, but the method has disadvantages. The final ordinations are the results of iteration and may not be globally optimal, depending on starting conditions. Thus, it is strongly recommended that any multidimensional scaling be repeated with different starting positions. If several different displays are produced, those with similarly low stress should have similar arrangements of the points, apart from an arbitrary left–right and up–down ambiguity. Multidimensional scaling with more than 25 or so individuals can take appreciable computer time, and may be impossible with more than 50 or a few hundred individuals (e.g., Fig. 5.4), depending on the computer package.

How can one tell whether a multidimensional scaling plot is "good"? An informal but useful criterion is that a stress of less than about 0.1 is an indicator of a useful ordination (Morgan et al. 1976). How many dimensions should be used? Two is easiest on the human eye–brain system, but using a three-dimensional plot (e.g., Fig. 5.6) or a series of plots of different dimensions against one another, we can assimilate information from more dimensions. Multidimensional scaling can be tried in a range of numbers of dimensions (e.g., two to five) and the optimal stress noted for each. These stress values should decrease as the number of dimensions increases. An analyst might choose the number of dimensions at which the stress first falls below 0.1, or that at which it starts to level out, so that additional dimensions make little improvement to the stress.

Table 5.2 Summary of Characteristics of Principal Coordinates Analysis and Nonmetric Multidimensional Scaling for Ordinating Matrices of Association Indices or Interaction Rates

	Principal coordinates analysis	Nonmetric multidimensional scaling
Scaling	Metric	Nonmetric
Matrix	Positive semidefinite (no negative eigenvalues)	–
Solution	Unique	Iterative, not necessarily optimal
Computing	Quick	Slow
Maximum individuals	Virtually unlimited	~50–200
Dimensions needed for "useful" ordination	More	Less
Choose number of dimensions	Afterward	Before
Measure of fit	Proportion of variance accounted for	Stress

Table 5.2 summarizes the relative advantages and disadvantages of nonmetric multidimensional scaling relative to principal coordinates analysis.

5.2.5: Cluster Analysis. Another method commonly used to display matrices of association indices or interaction rates is hierarchical cluster analysis (Section 2.7), in which the individuals and their relationships are displayed in a dendrogram, or tree diagram (Morgan et al. 1976; Fig. 2.4).

Dendrograms are attractive ways of displaying matrices of association indices or interaction rates, but unlike sociograms or the ordinations of principal coordinates analysis or multidimensional scaling, they impose a model. The model is that the society is structured in a hierarchical fashion, in the sense that clusters of individuals formed at the high-association end of the dendrogram, the groups of twigs, act as subjects in the social structure at the low-association end, the branches. Here, "hierarchical" is not used in the sense of a dominance hierarchy. A dendrogram may be a reasonable representation of a society, for instance, in the second-order alliances of first-order alliances formed by male bottlenose dolphins (*Tursiops* spp.) (Connor et al. 1992), or it may not, for instance, in the loose network of associations formed by female bottlenose dolphins (Smolker et al. 1992).

Random data can make interesting dendrograms. The dendrogram of associations among 11 individuals shown in Fig. 5.7 suggests a quite complex society. It is divided into two principal clusters, each containing some pairs with strong (association indices > 0.65) relationships. Are these mated pairs? But Fig. 5.7 is a dendrogram drawn entirely using randomly generated associations.

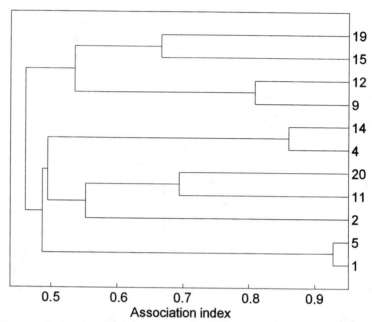

FIGURE 5.7 Dendrogram produced using average-linkage hierarchical cluster analysis on a random matrix of association indices (all association indices are the mean of two numbers chosen from the [0, 1] uniform distribution).

There is a way to weed out at least some of such spurious dendrograms: the cophenetic correlation coefficient (CCC). As noted in Section 2.7, the CCC is the correlation between the dyadic entries in the matrix of association indices and the level at which the dyads are joined on the dendrogram (Bridge 1993). A CCC of 1.0 indicates a perfect fit, so that the dendrogram is an appropriate model of the matrix of association indices or interaction rates, fully describing it. A reasonable rule of thumb is that a dendrogram is a reasonable display of a matrix if the CCC is greater than 0.8 (Bridge 1993). The dendrogram from the random data in Fig. 5.7 has a CCC of 0.58 and so is rejected under this criterion, whereas those from the real data in Figs. 5.2 to 5.4 are all acceptable. This does not mean they are truly hierarchically structured societies, but that a hierarchically structured model is reasonably consistent with the data.

The dendrogram of capuchin grooming data (Fig. 5.2) indicates a frequently grooming dyad, S and N, that forms the nucleus of the grooming network among these females, with no other preferred partnerships, and differences in grooming rates among individuals.

With 55 disk-winged bats, the dendrogram (Fig. 5.3) is becoming cluttered, but, perhaps with the help of a magnifying glass, an informative structure emerges. There are a many links with associations of 0.6 or higher, producing clusters of two to nine individuals that frequently roost together but rarely roost with other clusters. There are also a number of "loner" individuals without strong associations. There seems to be little structuring of the roosting society based on sex.

The dendrogram of hundreds of sperm whales (Fig. 5.4) is too full to be assimilated in anything except its coarsest features. With a CCC of 0.92, however, it is an accurate representation of the matrix of association indices. There are strongly associated clusters, each containing a dozen or so individuals, and very few loners. Because many of the links are at an association index of zero, there are effectively discrete communities of whales. The overcluttered dendrogram of Fig. 5.4 can then be dissected into more informative diagrams showing these closed communities. In Fig. 5.8, the sperm whale dendrogram is divided along the x-axis into six separate displays with almost no association between them. It becomes apparent that many of the unlinked animals are males, and some fairly clear, and apparently closed, social units of animals linked by high association indices appear.

Looking at such dendrograms, the social analyst may wonder where to place the dividing line along the "y" (association index) axis if she wants to retrieve the membership of the units. This is a topic for Section 5.7 on assigning individuals to social units and other entities.

Technically, hierarchical cluster analysis can be performed in a number of ways [for a rather complete description of clustering methods, see Legendre and Legendre (1998, pp. 314–355); for a shorter summary, see Manly (1994, pp. 128–134)]. Most techniques are agglomerative, starting with all individuals alone in their own clusters. Then, at each step, the most associated individuals/clusters are merged, building up the dendrogram. However, an important decision must be made: After clusters A and B are merged, what is the association of the new AB cluster with another cluster C? There are several ways that these associations can be calculated, some of which (such as Ward's method) depend on the associations representing distances in Euclidean space. This is not the case with matrices of association indices or interaction rates, although either can be transformed into Euclidean distances using principal coordinates analysis or multidimensional scaling (see prior discussion), or we can just assume that the square root of one minus the association index is a distance. Such steps seem unnecessary because there are perfectly good

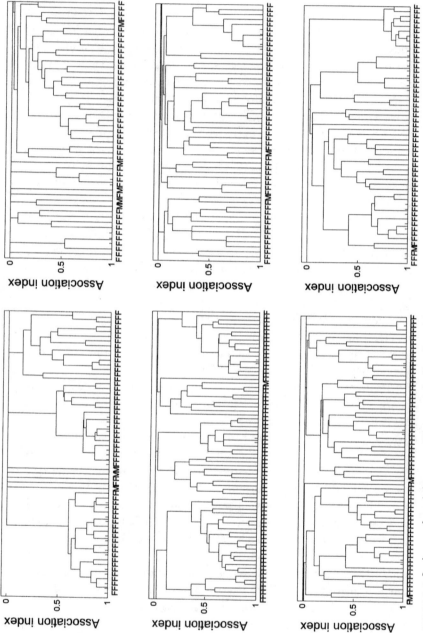

FIGURE 5.8 Dendrogram of associations among 353 sperm whales (from fig. 5.4) divided into six sections. F, female or immature; M, mature male.

clustering methods that work directly on similarity matrices such as matrices of association indices or interaction rates (although these must be symmetric). These include the following:

1. Single-linkage (nearest neighbor) cluster analysis. After clusters A and B are merged, the similarity between AB and C is the most similar of AC and BC.
2. Complete-linkage (furthest neighbor) cluster analysis. The similarity between AB and C is the least similar of AC and BC.
3. Average-linkage (or unweighted group mean) cluster analysis. The similarity between AB and C is the average of all similarities between individuals in C and either A or B.

Although Morgan et al. (1976) initially recommended single-linkage cluster analysis, more recent research has strongly indicated that average-linkage cluster analysis is most likely to mimic the input matrix (Milligan & Cooper 1987). This recommendation is perhaps especially warranted in social analysis because extreme small or large similarities, whether caused by random error, measurement error, or unusual relationships, have less impact on the results than in single-linkage or complete-linkage cluster analysis (Whitehead & Dufault 1999). The CCC can be used to indicate the best method: Choose the method with the highest CCC. In my trials, this has usually been average linkage.

There are other ways of producing dendrograms, such as hierarchical divisive methods in which all individuals are initially placed in the same cluster and this is sequentially divided up, perhaps using K-means (Section 2.7). This is in some ways a theoretically good method (Legendre & Legendre 1998, pp. 343–349) but is computationally intensive and not much used (Manly 1994, p. 132) outside of network analysis (Section 5.7).

5.2.6: Other Methods. A variant on the dendrogram produced by hierarchical cluster analysis is the additive tree (Sattath & Tversky 1977). Additive trees are a form of network representation. As in the traditional dendrogram, individuals are represented by nodes, but the dissimilarity between two individuals is estimated by the length of the path joining their nodes, and so the nodes representing the individuals are not necessarily aligned along a common axis. Additive trees are a good way of displaying dissimilarity matrices, especially when within-cluster dissimilarities are greater than within-cluster similarities, and they are much used in displaying phylogenetic relationships, whether derived from molecular genetic or other data. However, they are aimed at describing dissimilarity matrices, have no direct meaning for similarity measures, such as association indices or interaction rates, and so are little used in social analysis.

A related and simpler technique, the minimum/maximum spanning tree, is used in social analysis (Morgan et al. 1976). This is a variant on the sociogram and is also related to network analysis. Individuals are represented by points or nodes in a graph. A "tree" is constructed by drawing links between the points such that each point is visited by at least one link, no closed loops occur, and we can get from any point to any other point by moving along the links. The maximum spanning tree is that in which the sum of the associations for the dyadic relationships represented by the links is maximized. Thus, it represents the most important set of relationships for connecting the whole population. A minimum spanning tree is the equivalent for a dissimilarity measure (such as genetic distance). I cannot see any advantages of the maximum spanning tree over the sociogram, other than that it might look less cluttered. There is also the disadvantage that important relationships with high association indices or interaction rates may not be linked in a maximum spanning tree. For instance, consider the following similarity matrix (perhaps a matrix of association indices):

A			
B	0.84		
C	0.87	0.80	
D	0.10	0.12	0.22
	A	B	C	D

The maximum spanning tree will link A and C, A and B, and C and D, omitting the strong association between B and C, at the expense of the quite weak association between C and D.

A frequently useful and simple display is a histogram of group sizes. These can be of the group sizes experienced by outside observers or the typical group sizes experienced by the animals themselves (Section 3.4), and they can be calculated from association indices (Section 4.3) or from observations of groups. The latter does not require individual identification of animals. The distributions of group sizes obviously tells us something about the potential number of interactants an individual has at any time, but it can also indicate processes of group dynamics. Cohen (1971) explored the distributions of group sizes expected, given different models of group formation. Based on this approach, some models of group formation may be ruled out for a particular data set given the lack of fit between the observed and expected distribution of group sizes. However, this analytical approach omits some of the most informative social information—the identities of the individuals—and has largely been superseded by modeling individual-specific social data, as in lagged association rates (Section 5.5) and Bayesian delineation of social units (Section 5.7).

BOX 5.2 *Visual Displays of Social Structure: General Guidelines*

For any kind of population, histograms give useful information about the distribution of any measure and thus how relationships are distributed in the population. Separating information for classes of dyad, or logging the y-axis, may improve information content and interpretability, respectively.

When numbers of animals are small, all of the displays considered in Section 5.2 are possible and potentially useful. I particularly like sociograms because they display almost all of the data, can indicate the form of social structure (unlike histograms, which are usually ambiguous in this respect), and, unlike cluster analyses or ordinations using principal coordinates analysis, can retain and present information on asymmetric relationships. Principal coordinates, multidimensional scaling, and cluster analyses all come with measures of fit (proportion of variance accounted for, stress, and CCC, respectively), and may or may not provide useful representations of a small data set in an acceptable number of dimensions (usually two or three). With small data sets, one can try them all and different types of cluster analysis and multidimensional scaling, retaining those that both provide a good quantitative description of the data, as indicated by the measure of fit, and a useful description of the social structure. If the fit of the ordination to the data is good, then principal coordinates and multidimensional scaling should not be very misleading. Dendrograms provide a nice, hierarchically structured model of social structure, but this can be misleading, as illustrated by the dendrogram of random data shown in Fig. 5.7.

With large population sizes, containing more than about 100 individuals, the options are more limited. Histograms still work well, but sociograms become cluttered with more than 100 individuals (although the network drawing packages can greatly help in producing useful sociograms; Section 5.3). Multidimensional scaling may either be impossible or not produce a repeatable display with low stress. Principal coordinates will work with large data sets, but the display may have little utility (as in Fig. 5.4). Cluster analysis appears to be useful for the large sperm whale data set when enlarged (Fig. 5.8), but this will not always be the case. With large population sizes, the less visual but more analytical methods, such as lagged association rates (Section 5.5), become relatively more important.

5.3 Network Analysis

5.3.1: Introduction to Networks. In recent years, the informative and sophisticated set of techniques collectively known as network analysis has begun to be used for the analysis of nonhuman social systems (e.g., Lusseau 2003; Croft et al. 2004; Flack et al. 2006). Network analysis has its theoretical basis in the mathematical discipline of graph theory and is the subject of a huge literature from a variety of perspectives and scientific disciplines, including the work of a remarkable number of physicists. Recent general reviews of network analysis include those by Boccaletti et al. (2006), Newman (2003b), and (particularly accessible) Proulx et al. (2005). For more information on the application of network analyses to non-human social networks, see the reviews by Wey et al. (2008) and Krause et al. (2007), and Croft et al.'s (In press) book.

Network analysis envisages a social system as a *network*, a set of *nodes* (vertices or points)—usually individuals, but they could be higher levels of social structure, such as units or communities—connected by *edges* that indicate their interactions. Networks are similar to sociograms and dendrograms in that there is a graphical display with relationships being denoted by links rather than Euclidean position as in the ordination techniques of principal coordinates and multidimensional scaling. In fact, sociograms, dendrograms, spanning trees, and dominance hierarchies are special cases of the general concept of a graph network.

Newman (2003b) notes four general areas of network analysis:

1. Drawing and viewing networks
2. Analyzing statistical properties of networks
3. Modeling networks
4. Predicting network behavior

Predicting network behavior (e.g., predicting how, in general, the cultural behavior of a population depends on its social structure) is in its infancy in all fields (Newman 2003b) and is not covered here. Network modeling is quite well developed in some areas, such as the analysis of the World Wide Web, but it has hardly started with nonhuman social networks. Thus, in this section, I principally discuss the generation, description, and statistical analysis of networks.

Networks come in a variety of types. In the simple, standard, "undirected binary" network, there is one kind of node (i.e., no classification of individuals) and one type of edge, so that individuals are either connected or unconnected. A variant relevant to social analysis occurs when nodes come in discrete classes (such as males and females; Section 3.6). Edges may also be allocated to classes (e.g., "friendship", "animosity"), be "weighted" (so they might represent interaction rates or association indices), or "directed" (as with an asymmetric interaction such as grooming).

The great majority of network analyses in studies of human social systems, information systems, technological systems, and biological systems consider only directed or undirected binary networks (Newman 2003b, 2004; Proulx et al. 2005). This is a drawback of network analysis in its current state. Although the analyst may be able to draw a binary edge with certitude (she saw the two individuals grooming or playing; e.g., Flack et al. 2006), she can rarely be certain that two members of a community are not linked. An alternative to the binary associated/not associated (or interacted/not interacted) edge is to have a cutoff association index (or interaction rate). This is also unsatisfactory in the respect that it is arbitrary (Fig. 5.9 shows three very different network depictions of the same data set on guppies, using three different cutoffs to define edges; see Section 4.10) and too sharp: If the mean association index is 0.2, a pair with an index of 0.19 will not be linked and one with an index of 0.21 will be, even though there is no evidence that the two relationships are different either statistically or in reality. Weighted network analyses are the way around this, and these are being developed (Barrat et al. 2004; Newman 2004; Barthélemy et al. 2005; Boccaletti et al. 2006; Li et al. 2006). Croft et al. (2004, 2005) made a start at "analysis of weighted networks" for nonhuman vertebrates, although the tests that they perform are actually more like the Bejder et al. (1998) permutation test (Section 4.9), with less control in the structure of the permutations (see later discussion).

One important benefit of using network analysis for nonhuman social analyses is the sophistication of the network-drawing computational

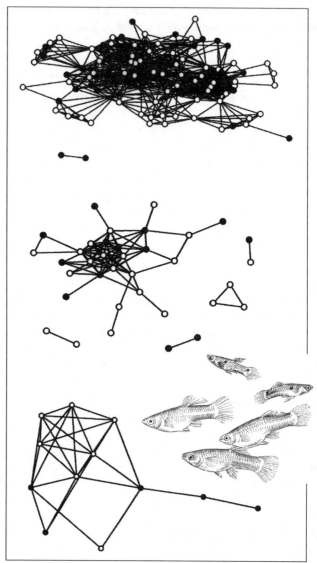

FIGURE 5.9 Network depiction (males, *filled circles*; females, *open circles*) of wild guppies (*Poecilia reticulata*) after 7 days of sampling with links between fish caught together in the same school (**upper**) at least once, (**middle**) at least twice, and (**lower**) at least three times. (From Croft et al. 2004, fig. 1a.) (Illustration copyright Emese Kazár.)

routines. In contrast to the rather basic routine in SOCPROG used to draw the sociograms in Figs. 5.1 to 5.4, which becomes quite uninformative with a few tens of individuals, the much more developed network software (e.g., in NetDraw; Section 2.9) gives useful representations with more than 100 (Fig. 5.9) or even thousands of nodes (Fig. 2 of Newman 2003b). Some network analysis programs (including NetDraw) can make the widths of the edges proportional to their weighting, producing sociograms as described in Section 5.2.

A number of measures are used to describe networks. In the next subsections, I discuss some of the most common, beginning with those that refer to dyads, moving on to those that are properties of nodes, and then considering statistics of entire networks. Where appropriate, I give the meaning of the term for binary networks and then consider whether it can be generalized to weighted networks (such as association indices or interaction rates) or directed networks (in which the relationship from A to B is not necessarily that from B to A).

5.3.2: Network Measures of Dyads

- *Edges and weights*. These dyadic measures are the fundamental network data. They note whether two nodes are linked by an edge or not in a binary network or, in a weighted network, the importance of the link (e.g., the value of the association index). Edges and weights can be directed or undirected.
- *Geodesic path length*. The geodesic path length between any pair of individuals is the fewest number of links joining them in a binary network. It is not defined, or is infinite, if two individuals are not linked. The geodesic path length of a dyad within a binary network is, in some circumstances, a good indicator of its relationship. If the geodesic path length is 1, then they are directly linked; if it is 6, the connection seems very tenuous. In weighted networks, path lengths can be calculated by making the "length" of any edge the inverse of its weight, and the geodesic path length of any dyad is the one that minimizes the sum of the lengths of the edges forming a route that joins them (Newman 2004; Holme et al. 2007). An alternative for association indices might be to use the square root of one minus the association index, as in principal coordinates analysis (Section 5.2). The relative benefits of different

types of transformation for weighted networks have not been examined.

- *Vertex similarity.* There are several ways in which the similarity of nodes can be compared. All of those that I know of [listed by Leicht et al. (2006)] refer only to binary networks. One set of measures (structural equivalence) refers to the similarity of neighbors: Pairs of individuals with the same set of neighbors have high structural equivalence, or vertex similarity in this sense. Leicht et al. (2006) derive measures of structural similarity that indicate how well a pair of nodes is linked.

5.3.3: Network Measures of Nodes

- *Class and attribute.* Nodes can be allocated to classes, such as male/female, or possess continuous attributes such as age (Section 3.6).
- *Degree and strength.* The *degree* of a node in a binary network is the number of edges connected to the node (Newman 2003b). For a weighted network, the corresponding property is or the sum of the weights on the edges connected to the node. Some (e.g., Barthélemy et al. 2005) call this the *strength* of a node, and I follow this terminology. For a weighted network of association indices, the strength is the individual's gregariousness (Section 4.3). For a directed network, representing an asymmetric relationship measure, each individual has an "in-degree" or "in-strength" (e.g., overall rate of being groomed) and "out-degree" or "out-strength" (e.g., overall rate of grooming).
- *Betweenness centrality.* The betweenness centrality—usually just called betweenness—of a node (Freeman 1979) in a binary network is the number of shortest paths between other nodes that pass through that node. Thus, peripheral individuals, that is, individuals only connected to one other individual, have zero betweenness, and an individual that links two otherwise discrete clusters has a very high betweenness. When a network is being used to investigate the flow of information or disease through a population, then betweenness is a very useful measure (Newman 2003b).

Individuals with high betweenness can have a large effect on the spread of the information or disease. If the network is simply being used to describe the social structure, however, then betweenness is less useful. For instance, if the population is divided into a number of fairly discrete communities, then the animals with greatest betweenness might be society's "social glue" (e.g., Lusseau & Newman 2004) or, alternatively and perhaps more parsimoniously, the socially exploratory young animals whose relationships have low impact on the nature or stability of the social structure. In weighted networks, betweenness centrality could be calculated similarly but using the sum of the inverses (or some other transform) of the weight on each edge to find the shortest path, as in geodesic path length (Newman 2004).

· *Eigenvector centrality.* Eigenvector centrality (Newman 2004) is another measure of how well connected an individual is within the network. Mathematically, it is simply the appropriate element of the first eigenvector of the matrix of edges or weights (e.g., a matrix of association indices). For each individual, it gives a number that indicates its connectedness within the network, so that an individual can have high eigenvector centrality either because it has high degree or strength or because it is connected to other individuals of high degree or strength. Although eigenvector centrality has a less direct definition than betweenness, it has the major advantages that it is directly available for weighted, nonbinary networks (Newman 2004) and does not presuppose the flow of some attribute, such as disease or information, through the network. Eigenvector centralities may be highly skewed, being high for a few well-connected individuals and virtually zero for the remaining members of the population.

· *Reach.* Reach is a measure of indirect connectedness, which Flack et al. (2006) defined, for a binary network, as the number of nodes two or fewer steps away. Other definitions might be used in other circumstances. For instance, in a weighted network of association indices, the reach of any individual A might be defined as the sum over other individuals B of the sum over other individuals C

Table 5.3 Suggested Nodal (Individual) Network Measures as Used on Weighted Networks Such as Matrices of Association Indices or Interaction Rates

Measure	What it means	Formula for weighted network
Strength (degree)	How connected to other individuals (equivalent to: gregariousness; mean typical group size minus 1)	$s_I = \sum_j a_{IJ}$
Eigenvalue centrality	How well connected, in terms of number and strength of connections, and to whom	$e_I = (\text{first eigenvector of } a)_I$
Reach	Overall strength of neighbors	$r_I = \sum_J a_{IJ} \cdot s_j$
Affinity	Weighted mean strength of neighbors	$f_I = r_I / s_I$
Clustering coefficient	How well connected neighbors are to each other	$c_{IJ} = \dfrac{\sum_J \sum_K a_{IJ} \cdot a_{IK} \cdot a_{JK}}{\max(a_{JK}) \cdot \sum_J \sum_K a_{IJ} \cdot a_{JK}}$

a_{IJ} is the association index or interaction rate between individuals I and J; $a_{II} = 0$ for all I. Formulas are from Newman (2004) and Holme et al. (2007).

of the products of all pairs of association indices linking A and B through another individual C (and this is the definition I use in the examples; see Table 5.3). Reach is a sensible and useful measure for a society that exhibits behavioral contagion (Flack et al. 2006): The dyadic behavior of A with B may induce dyadic behavior between B and C.

· *Affinity.* The affinity of a node in a binary network is the average degree of its neighbors (e.g., Barthélemy et al. 2005). Thus, a node with high affinity is connected to other nodes of high degree. Barthélemy et al. (2005) and Barrat et al. (2004) suggest generalizing affinity to weighted networks using the weighted degree of the neighbors, where the weight is that of the edge joining the focal individual to its neighbor. I believe that the average weighted strength of the neighbors is a more useful measure of affinity in nonhuman social systems, making the affinity of a node its reach divided by its strength (Table 5.3).

· *Clustering coefficient.* The clustering coefficient (called the "network density" in sociology; Newman 2003b) measures

the degree to which the associates of an individual are themselves associated in a binary network. If all associates of an
individual are themselves linked, then the clustering coefficient of that individual is 1.0; if none of them is, it is 0.0.
Clustering coefficients are high in societies containing
tight, closed, homogeneous social units and are lower in
strict territorial societies in which individuals only associate with their neighbors, who may not associate with
each other. The clustering coefficient seems to be a useful
measure of individual sociality (Croft et al. 2004) and is
widely used in sociology (Newman 2003b). Generalizations of clustering coefficients to weighted networks have
being developed (Barthélemy et al. 2005; Li et al. 2006).
The most satisfactory clustering coefficient for use on matrices of association indices seems to be that of Holme
et al. (2007), which considers the weight on all three edges
of each triangle linking nodes, and this is the definition
that I use (Table 5.3). This is standardized by the maximum weight in the network, however, and so it is heavily
influenced by, say, two inseparable animals, and thus is
hard to compare between populations or even between
a social network and its random permutations (see later
discussion).

· *Disparity*. This is a measure of the variance in the weights
of the edges connecting a node in a weighted network
(Barthélemy et al. 2005), and so it appears to be a useful
measure of the sociality of an individual:

$$Y(I) = \sum_J \left(\frac{\alpha_{IJ}}{s_I} \right)^2 \qquad (15)$$

Here, s_I is the gregariousness of I and α_{IJ} is the association
index between individuals I and J. An animal with high disparity has varied social relations, whereas one with low disparity has a much more homogeneous set of social partners.
It is thus an individualized alternative to social differentiation introduced in Section 5.1. The variance of association
indices or interaction rates depends heavily on the size and
structure of the sample (Sections 4.4 and 4.5), however, and
so disparity measures calculated from such data will only be

representative if sample sizes are very large or corrections for sampling are introduced (I do not know of any attempts to do this).

The most useful of these nodal measures for the analysis of association indices or interaction rates, for which we need to consider weighted networks, are class, strength, eigenvector centrality, reach, affinity, and the clustering coefficient. Table 5.3 gives suggested mathematical formulas.

5.3.4: Network Measures for a Population or Community. We can take the means of any of the dyadic or nodal measures introduced previously to describe a network. So we have the mean geodesic path length, mean degree, or mean clustering coefficient as network descriptors (e.g., Lusseau 2003; Newman 2003b; Croft et al. 2004). Some care must be taken for geodesic path length, which is not defined for unconnected dyads, and clustering coefficient and disparity, which are not defined for isolated individuals. Confidence intervals for mean network measures can be calculated using the bootstrap method (Section 2.3) in which random replicates are constructed by choosing sampling periods with replacement (Lusseau et al. In press).

Other summary statistics of the network measures can be useful. Standard deviations or coefficients of variation indicate how a measure varies among nodes or dyads, but these also include sampling variation. As another example, the diameter of a network is (usually) the maximum (across dyads) geodesic path length (Newman 2003b).

Distributions of the measures are more informative than summary statistics. Distributions and summary statistics can be compared with the distributions of the statistics expected theoretically. There has been considerable mathematical work on the expected distributions of network statistics for artificial networks created using different algorithms (Newman 2003b). For instance, networks generated by preferential-attachment models, in which well-connected nodes are more likely to become linked, have a power-law-degree distribution that is known as "scale free" (Barabási & Albert 1999).

In social analysis, it is probably more useful to compare distributions or summary statistics of network measures with those calculated from randomly constructed networks using the same individuals and data characteristics (e.g., Croft et al. 2005). The methods described in Section 4.9 are useful for producing random networks subject to sampling and gregariousness constraints. Thus, the original Bejder et al.

(1998) method can produce random networks subject to the sizes of all groups observed and the number of groups in which each individual was observed, and using the modifications described by Whitehead (1999a; Section 4.9), we can control for each individual's number of groups or associations in each sampling period.

Other attributes of a social system can be measured by comparing different nodal measures. For instance, degree or strength and affinity (the mean degree or strength of an individual's neighbors) may be related. If nodes with high degree/strength also have high affinity, then the "important" individuals are preferentially linked with each other, which is known as *assortative* mixing, a situation that seems generally characteristic of social networks (Newman 2003b). In disassortative mixing, which is not usually found in social networks but is present in other networks (e.g., the internet; Newman 2003b), the nodes of highest degree are surrounded by low-degree nodes. Thus, assortativity is potentially a useful descriptor of social organization. We can measure assortativity in a binary network simply by correlating the degrees of the individuals at each end of an edge (Newman 2003b), using the standard correlation coefficient. Flack et al. (2006) showed that the experimental removal of three high-status "policing" macaques (*Macaca nemestrina*) from a colony reduced this measure of assortativity. For weighted networks, the correlation between the strength of a node and its affinity is a suitable alternative (Barthélemy et al. 2005).

When nodes are divided into classes by gender, age, or other attributes (Section 3.6), then the network measures can be calculated for each class and compared, such as the mean strength, connectivity, or reach of males versus females. We can also calculate within-class values for all measures (such as the strengths of the nodes in the all-female network) and between-class measures for most (the mean female–male strength would be the mean, over females, of the sum of the weights along edges leading from each female to all males, as with interclass gregariousness; Section 4.11).

With classes defined, another type of assortativity can be assessed. Here, one measures the degree, or strength, of the interactions between and within classes and compares them. There are several possible assortativity coefficients. The best seems to be that of Newman (2003a), which is defined using matrix algebra. It equals 0 in a randomly mixed network, in which there is no preference or avoidance between classes, 1 when all edges join members of the same class, and -1 when all edges join members of different classes. As far as I can tell, it has not been generalized for weighted networks, although the matrix correlation

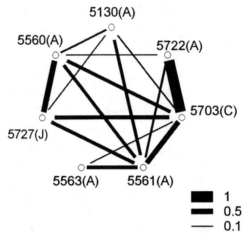

FIGURE 5.10 Sociogram of matrix of association indices of the group of seven sperm whales in table 2.5 (Gero 2005). A, adult female; C, male calf; J, juvenile male.

coefficient and Mantel test between a matrix of association indices or interaction rates and the 1:0 (same: different) matrix of same/different class (Section 4.11) provide good analogs. Nodal network measures can also be compared with continuous individual attributes. For instance, we could plot connectivity against age, which might, for instance, indicate that individuals form clearer cliques as they become older.

Network analysis also includes routines for finding "clusters" of nodes within networks, which are well connected internally but have few links between them. Such clusters might correspond with my concept of the social unit or community (Section 1.6). Some of these routines are described in Section 5.7. Once a network has been dissected using these methods or non–network-based methods of assigning individuals to units (Section 5.7), network statistics can be calculated for any or all clusters separately or between clusters.

5.3.5: Two Examples. To provide a feel for the power and limitations of network analysis, here are two examples. In these examples, the definitions used for the network measures are as described in the preceding subsections and in Table 5.3.

For a first and simple example, consider the social unit of seven sperm whales whose matrix of association indices is shown in Table 2.5. Figure 5.10 shows a sociogram, or network, of their association indices, and Table 5.4 lists five weighted network measures. Both the sociogram and

Table 5.4 Network Measures (Standard Errors from 1,000 Bootstrap Replicates in Parentheses) for Half-Weight Association Indices between Members of a Group of Seven Sperm Whales (*Physeter macrocephalus*) Studied off Dominica in 2005

ID	Strength	Eigenvector centrality	Reach	Clustering coefficient	Affinity
5703(C)	2.36 (0.22)	0.55 (0.02)	3.43 (0.57)	0.18 (0.03)	1.45 (0.12)
5722(A)Mother of 5703	1.43 (0.12)	0.43 (0.04)	2.96 (0.38)	0.33 (0.06)	2.07 (0.18)
	[0.46 (0.12)]	[0.21 (0.05)]	[0.51 (0.16)]	[0.51 (0.12)]	[1.13 (0.15)]
5561(A)	1.87 (0.23)	0.42 (0.03)	2.74 (0.57)	0.23 (0.03)	1.47 (0.15)
	[1.39 (0.18)]	[0.52 (0.03)]	[1.18 (0.26)]	[0.27 (0.05)]	[0.85 (0.10)]
5727(J)	1.45 (0.21)	0.36 (0.04)	2.46 (0.53)	0.30 (0.05)	1.70 (0.17)
	[1.10 (0.16)]	[0.50 (0.04)]	[1.24 (0.27)]	[0.42 (0.10)]	[1.13 (0.12)]
5560(A)	1.49 (0.21)	0.36 (0.04)	2.39 (0.52)	0.28 (0.05)	1.60 (0.17)
	[1.20 (0.16)]	[0.51 (0.04)]	[1.22 (0.25)]	[0.36 (0.09)]	[1.01 (0.11)]
5130(A)	0.82 (0.22)	0.20 (0.05)	1.37 (0.42)	0.33 (0.05)	1.68 (0.18)
	[0.68 (0.19)]	[0.31 (0.08)]	[0.79 (0.25)]	[0.52 (0.10)]	[1.15 (0.14)]
5563(A)	0.76 (0.19)	0.19 (0.04)	1.31 (0.44)	0.33 (0.05)	1.72 (0.18)
	[0.63 (0.16)]	[0.28 (0.06)]	[0.72 (0.24)]	[0.46 (0.12)]	[1.14 (0.15)]
Means	1.45 (0.56)	0.36 (0.13)	2.38 (0.79)	0.28 (0.06)	1.67 (0.21)
	[0.91 (0.37)]	[0.39 (0.14)]	[0.94 (0.31)]	[0.42 (0.09)]	[1.07 (0.12)]

A, adult female; C, male calf; J, juvenile male). Values in square brackets are for the situation after the removal of the calf from the data set (a "topological knockout"). Data from table 2.5.

network analyses indicate the rather peripheral positions of #5130 and #5563 and emphasize the central role of the calf: The calf has much the highest strength, eigenvector centrality, and reach but the lowest clustering coefficient and affinity. In this network, rather unusually (Newman 2003b), strength is negatively related to both the clustering coefficient ($r = -0.92$) and affinity ($r = -0.48$).

Removing the calf—a "topological knockout" (Flack et al. 2006)—changes the structure of the network. For instance, #5722, the mother of the calf, has the largest reach of the adults when the calf is in the network and the smallest reach when it is removed: She interacts with the unit principally through the calf. In addition, after the calf is removed, mean strength, reach, and affinity are reduced, whereas clustering coefficient is increased, presumably because the relationships are now more homogeneous.

For a more complex example, consider the 55 bats (individuals identified on 5 or more days) whose network of co-roosting relationships is shown in the sociogram of Fig. 5.5. Table 5.5 summarizes the network measures for this population. All of the network measures vary considerably between individuals (indicated by high standard deviations relative to the mean), although how much of this is due to sampling variation is unclear from the straight means and standard deviations. The results indicate a society mildly structured by gender, with males having somewhat higher values of all the network statistics, greater strength, reach, centrality, clustering coefficients, and affinity. However, there is considerable variance and overlap between the sexes. For instance, the standard errors of the network measures (calculated using 1,000 bootstrap replicates, choosing sampling periods randomly with replacement) suggest little real difference between the sexes in any measure (because the difference between the sexes was less than twice the sum of the standard errors for each sex; Table 5.5), except eigenvector centrality. Between-sex strength is somewhat greater than within-sex strength, indicating a slightly assortative society. This is confirmed by the matrix correlation coefficient between the association indices and the 1:0 matrix of same/different gender ($r = -0.035$; $P = 0.09$, two-sided Mantel test).

We can also look at classes defined by the clear clusters in Fig. 5.5 (the same clusters are produced by an average-linkage cluster analysis with a cutoff average association index between clusters of 0.13, which maximizes modularity; see Section 5.7). Generally, larger clusters have higher values of all the network measures, although there are some startling

Table 5.5 Mean Network Measures for Half-Weight Association Indices between 55 Co-roosting Disk-winged Bats (Thyroptera tricolor), [a] Overall, between and within Genders, and between Clusters Containing More Than Two Individuals (as shown in fig. 5.5)

	Strength	Eigenvector centrality	Reach	Clustering coefficient	Affinity
Overall	2.86 (1.90) SE = 0.19	0.05 (0.12) SE = 0.00	11.73 (12.96) SE = 1.78	0.67 (0.15) SE = 0.05	3.21 (1.69) SE = 0.21
Among sex classes					
Males-all ($n = 27$)	3.16 (1.91) SE = 0.25	0.07 (0.14) SE = 0.00	13.91 (13.94) SE = 2.37	0.69 (0.16) SE = 0.05	3.48 (1.78) SE = 0.25
Females-all ($n = 28$)	2.58 (1.88) SE = 0.15	0.04 (0.11) SE = 0.00	9.62 (11.81) SE = 1.26	0.64 (0.14) SE = 0.05	2.91 (1.57) SE = 0.18
Male-male	1.49 (1.35)	0.09 (0.17)	3.96 (5.43)	0.63 (0.19)	1.76 (1.29)
Male-female	1.67 (0.72)	–	–	–	–
Female-male	1.61 (1.41)	–	–	–	–
Female-female	0.97 (0.66)	0.07 (0.18)	1.36 (1.26)	0.67 (0.21)	1.26 (0.42)
Among clusters of bats					
A-all ($n = 6$)	2.74 (0.91)	0.00 (0.00)	8.35 (2.94)	0.62 (0.23)	3.01 (0.15)
B-all ($n = 9$)	6.24 (0.45)	0.33 (0.02)	39.13 (2.57)	0.79 (0.01)	6.27 (0.05)
C-all ($n = 7$)	3.54 (0.30)	0.00 (0.00)	12.62 (0.91)	0.59 (0.02)	3.57 (0.05)
D-all ($n = 3$)	0.94 (0.17)	0.00 (0.00)	0.91 (0.11)	0.47 (0.17)	0.97 (0.06)
E-all ($n = 5$)	3.08 (0.32)	0.00 (0.00)	9.55 (0.91)	0.77 (0.03)	3.11 (0.04)
F-all ($n = 7$)	2.48 (0.66)	0.00 (0.00)	6.44 (2.49)	0.60 (0.22)	2.52 (0.42)
G-all ($n = 4$)	1.78 (0.37)	0.00 (0.00)	3.28 (0.58)	0.63 (0.10)	1.85 (0.07)
H-all ($n = 5$)	2.93 (0.32)	0.00 (0.00)	8.68 (0.82)	0.74 (0.04)	2.97 (0.05)

[a] See Vonhof et al. (2004); just using individuals identified on 5 or more days. Standard deviations of individuals are given in parentheses, and estimated standard errors of the means, calculated using 1,000 bootstrap replicates of sampling periods, are listed in some cases.

differences. Most obviously, the eigenvector centrality is zero (to two decimal places) for all clusters except cluster B, the most strongly connected. Thus, in cases like this, eigenvector centrality is heavily loaded on the most connected clusters. Reach also shows pronounced variance among clusters. In a highly segmented society like this, the mean affinity of a cluster is almost identical to its mean strength, and there tend to be strong correlations over individuals between measures. In the bat data, strength and affinity are correlated at $r = 0.97$ and strength and clustering coefficient at $r = 0.49$.

Table 5.6 shows the results of some permutation tests on this data set. I calculated three network measures: the standard deviation of the strengths and the correlation coefficients over individuals between strength and clustering coefficient and between strength and affinity for the real data and for 1,000 permutations (with 100 flips per permutation where appropriate) of each of four types: (1) the simple method used by Croft et al. (2005), in which the individuals identified in each roost were randomly chosen from the entire population; (2) the method of Bejder et al. (1998; Section 4.9), in which the identities are permuted but roost sizes and numbers of observations of each individual are both kept constant; (3) the modification (Whitehead 1999a; Section 4.9) in which roost sizes and numbers of observations of each individual in each sampling period are kept constant; and (4) the version in which the number of associations of each individual in each sampling period is kept constant (Whitehead 1999a; Section 4.9). The expected values of the measures (the means over the 1,000 permutations) varied considerably among the permutation types, as did the significance levels (indicated by the proportion of values of the measure from the random permutations more extreme than the real value). Remarkably, in the case of the correlation between strength and clustering coefficient, the real value was either significantly large or significantly small, depending on the type of permutation used (Table 5.6). To interpret these tests, we need to consider carefully the structure of the data. In the case of the disk-winged bats, these discrepancies can be partially explained by the facts that the study area was divided into three subareas (indicated in Fig. 5.5) with nonoverlapping populations and that on each sampling period (day), only one subarea was sampled (Vonhof et al. 2004). Thus, the first two permutation types mix animals from different subareas, whereas the second two retain the distinctiveness of the subareas, very much changing the nature of the random networks.

Table 5.6 Three Network Measures for Disk-winged Bat Data Compared with Expected Values from 1,000 Random Permutations, Together with Two-Sided *P*-Values from Permutation Tests, for Four Types of Permutations

	SD(strength)	r (strength, clustering coefficient)	r (strength, affinity)
Real value	1.900	0.489	0.969
Permutations controlling for:			
Number of animals in each roost	0.740 ($P < 0.001$)	−0.308 ($P < 0.001$)	0.184 ($P < 0.001$)
Number of animals in each roost and number of observations for each animal	0.672 ($P < 0.001$)	−0.121 ($P = 0.010$)	0.267 ($P < 0.001$)
Number of animals in each roost and number of observations for each animal in each sampling period (day)	1.111 ($P < 0.001$)	0.566 ($P = 0.456$)	0.676 ($P < 0.001$)
Number of associates (sharing roost) of each animal in each sampling period (day)	1.769 ($P < 0.001$)	0.751 ($P = 0.022$)	0.875 ($P < 0.001$)

The four types of permutations are the randomization of identities for given group sizes [as in Croft et al. (2005)]; randomization of identities fixing group sizes and the number of observations of each animal (Bejder et al. 1998; section 4.9); randomization of identities fixing group sizes and the number of observations of each animal in each sampling period (Whitehead 1999a; section 4.9); and randomization of associations in each sampling period, given the number of associations of each individual in each period (Whitehead 1999a; section 4.9).

BOX 5.3 *Network Analyses: Recommendations*

Network analysis is much more vibrant, broader, and faster evolving than other areas of nonhuman social analysis. The physicists and mathematicians driving the field keep coming up with new ways of examining networks. Some give us insight into animal societies, whereas others do not, and the relevance of a particular technique or measure very much depends on the situation. It is an exciting area, but also a confusing one. There are many possibilities, and some of those used in the first network analyses of nonhuman societies were not optimal. Here are some general recommendations drawn from my initial exploration of the possibilities of network analysis (see also Lusseau et al. In press). This recommendation set is conservative, in the sense that it is unlikely that a researcher who follows them will reach very erroneous conclusions, but she may miss some excellent opportunities. The field is in flux, and the recommendations will likely change, principally becoming wider as social analysts gain experience with network analysis (e.g., a measure of disparity corrected for sampling effort might prove very useful). Here are my recommendations:

- Make good use of the network drawing routines, such as NetDraw. They give flexible and very useful views of societies.
- Almost always use weighted network measures, especially for association indices (Lusseau et al. In press) and interaction rates, rather than binary network measures. An analyst would need to have much more data than are usually available to make a o edge in a binary network reliable, and the differences among the edges marked by 1's is lost.
- Recognize that some network measures are closely related to more traditional concepts in social analysis. For instance, when association indices are used, the strength of an individual is simply its gregariousness or typical group size minus one (Section 4.3).
- Recognize that, in some cases, more traditional methods may achieve a particular goal more easily or effectively than network analyses. Assortativity in

a binary node characteristic such as gender may be best measured and tested using the matrix correlation coefficient between the association indices and the 1:0 matrix of same/different gender and a Mantel test (Section 4.11, Schnell et al. 1985).

· Consider carefully the rationale for using a particular measure. For instance, the dyadic measure betweenness only makes much sense if something (such as information or disease) is being transmitted along the edges of the network, and the measure reach becomes important in a society exhibiting contagion, so that if an individual is engaged in a dyadic interaction with one individual, it makes it more likely that it will interact with another (Flack et al. 2006).

· Comparisons within networks are usually quite straightforward and informative: identifying the individuals with greater strength or connectivity, or comparing these measures between males and females. Standard errors of the measures should be presented so that the reliability to be attached to these within-network comparisons can be assessed. These can be estimated using the bootstrap method (Section 2.3).

· Comparisons between networks are more fraught. Network measures such as the standard deviation of strength or the correlation between strength and connectivity depend greatly on factors such as the sampling scheme, sampling period, and association index used, as well as on the structure of the data (Croft et al. 2005; Table 5.6). Whereas elegant and informative methods of comparing binary networks have been developed (e.g., Faust & Skvoretz 2002), the comparison of weighted networks is more complex and much less developed.

· Tests of network measures against the values expected from null models (e.g., Croft et al. 2005) can be informative but must be carefully considered. These are best done using permutation tests, but I strongly recommend that the permutations be controlled for both the sizes of groups [as proposed by, and using the methods of, Bejder et al. (1998);

Section 4.9] and perhaps the gregariousness of individuals (Whitehead 1999a; Section 4.9), and in
data sets that have important temporal or spatial
structure, permutations should be carried out within
sampling periods (Whitehead 1999a; Section 4.9).
The results of the tests may be quite different, depending on how the permutations are controlled
(e.g., Table 5.6). It is also important to select the
test statistic carefully (see Section 4.9 for some
considerations) and consider clearly what the null
and alternative hypotheses mean in social terms.

In this section, I investigated the potentiality of network analysis from the perspective of association indices that are bound
by 0 and 1. I suspect that almost all the suggested analyses will
also work with symmetric interaction rates. The network analysis
possibilities with asymmetric interaction rates are more limited
but at least include strength (in-strength and out-strength).

5.4 Dominance Hierarchies

A particular, and important, type of social network is the *dominance hierarchy*. In many cases, the dominance hierarchy is not only a useful descriptor of social dynamics, it also is the way in which the animals themselves experience, and probably envision, a large part of their social
world: These individuals are to be dominated, those are to be appeased
(Drews 1993). Dominance hierarchies are described using one or more
strongly asymmetric interaction measures, such as the winners of agonistic encounters, submissive behavior, or priority access to resources
(Section 4.8). Ideally, the nodes of the dominance network—the individuals—can be arranged linearly from the most to the least dominant
individual with all directed edges pointing in one direction. When the
nodes are arranged to best satisfy this ideal, any edge pointing in the
wrong direction is termed an *inconsistency* (sometimes called a "dominance reversal," although this can also refer to a change in dominance
direction over time). An arrangement without inconsistencies—a perfect linear hierarchy—is sometimes achievable (e.g., Table 5.7(i), and
when it is, little further analysis is needed. As a caution, Appleby (1983)
points out that with a binary 1:0 asymmetric measure of dyadic dominance and five or fewer individuals, the individuals can be arranged in

Table 5.7 Dominance Relationships in a Group of Young Roosters (*Gallus domesticus*) at (i) 32 Weeks of Age and (ii) 16 Weeks of Age

		Subordinate											
	(i)	YY	B	G	R	W	Y	(ii)	YY	B	G	R	W	Y
	YY	–	1	1	1	1	1	YY	–	1	1	1	1	1
	B	0	–	1	1	1	1	B	0	–	0	1	1	1
Dominant	G	0	0	–	1	1	1	G	0	1	–	1	0	1
	R	0	0	0	–	1	1	R	0	0	0	–	1	1
	W	0	0	0	0	–	1	W	0	0	1	0	–	1
	Y	0	0	0	0	0	–	Y	0	0	0	0	0	–

From Murchison (1935).

a perfect linear hierarchy with probability greater than 0.05 even when the dyadic dominance relations are randomly assigned.

5.4.1: Measures and Tests of Linearity and Steepness. In many, perhaps most, cases, there are inconsistencies (there are two in Table 5.7(ii), and further analysis may be useful. A frequently used measure of dominance *linearity* in a society is Landau's (1951) index:

$$h = \frac{12}{n^3 - n} \cdot \sum_{I=1}^{n} (v_I - (n-1)/2)^2 \qquad (16)$$

where n is the number of animals in the population and v_I is the number of animals dominated by individual I (plus half of the number of animals with whom I does not have a dominance relationship, if relevant). h ranges from 0.0 in the completely nonhierarchical situation in which each animal dominates exactly half of the others [as, in this case, $v_I = (n-1)/2$ for all I] to 1.0 in the case of a completely linear dominance hierarchy without inconsistencies. Under the null hypothesis of random dominance relationships, the expected value of h is equal to $3/(n+1)$. Values of h greater than about 0.9 are usually taken as indicating a nearly linear hierarchy (Lehner 1998, p. 333). For the data in Table 5.7, $h = 0.77$ for the young roosters (Table 5.7ii), and $h = 1.00$ for the perfectly linear hierarchy of the older animals (Table 5.7i).

Appleby (1983) shows how to use h as a test statistic of the null hypothesis that dyadic dominance relations are random rather than linear, and de Vries (1995) generalizes h and Appleby's (1983) tests to deal with unknown dyadic dominance relationships (maybe the pair was never observed to interact). de Vries' (1995) h' is given by

$$h' = h + 6u/(n^3 - n)$$

where u is the number of unknown dominance relationships and h is the value of Landau's h for the original matrix with all dyadic entries whose dominance status is unknown being assigned the value 0.5. de Vries' (1995) test of linearity is a randomization test (Section 2.4).

Another potentially useful summary statistic of a dominance hierarchy is de Vries et al.'s (2006) *steepness*. The steepness of a hierarchy describes how certainly a dominant wins an interaction over a subordinate. In a very steep hierarchy (steepness ~1.0), dominants almost always win over subordinates, whereas in a shallow one (steepness ~0.0), results are unpredictable. More linear hierarchies tend to be steeper, but linearity and steepness are different measures. A shallow but linear hierarchy would be one in which, in all dyads, the higher-ranking individual is more likely to win a contest but not much more likely. Thus, steepness quantifies the egalitarian–despotic continuum (de Vries et al. 2006). Steepness is calculated by regressing a normalized version of David's dominance index (see later discussion) on rank. de Vries et al. (2006) also show how to use randomization tests to test whether the steepness is significantly greater than would be expected if the number of interactions won in each dyad was random.

5.4.2: Dominance Ranks and Dominance Indices. Considerable attention has focused on assigning individuals in a dominance hierarchy to a dominance rank (i.e., a unique integer between 1 and n, the number of animals in the population) or a noninteger dominance index. Lehner (1998, pp. 336–338), de Vries (1998), and Bayly et al. (2006) summarize some of the proposed methods, although there are important recently developed techniques that are not described in any of these reviews (see later discussion). The two general approaches have different ostensible objectives but are linked. If each individual is given a dominance index, then the individuals can be ranked using it, and if individuals are ranked, then a simple index is the rank divided by the number of individuals. Some methods use only binary dominant/subordinate information for each dyad, whereas others consider the data used to come to this conclusion (e.g., numbers of wins or losses).

It is worth noting that if the hierarchy is perfectly linear (i.e., $h = 1.0$), then assigning dominance ranks is trivial, and if it is far from linear (i.e., $h << 1.0$) or tests (Appleby 1983; de Vries 1995) do not reject the null hypothesis of random dyadic dominance, then it probably should not be attempted.

Here are some of the more popular methods for constructing hierarchies, beginning with the ranking methods:

- The "I" method (Slater 1961). Find a ranking that minimizes the number of inconsistencies. This seems simple and intuitive, but it might not produce a unique hierarchy of individuals because several arrangements of the individuals may possess the same, minimum, number of inconsistencies. For a few individuals, the I ordering can be found by hand, but with larger numbers, an iterative computerized procedure is needed (de Vries 1998).

- Find the ranking that minimizes the proportion of interactions in which an individual lower in the ranking wins over a more highly ranked individual (Brown 1975, p. 86). This method is vulnerable to different interaction rates between different dyads because an inconsistency involving a large number of interactions will be overweighted.

- The "I&SI" method (de Vries 1998). This is similar to the I method, but parts of the hierarchy that are unresolved by minimizing the number of inconsistencies (the same number of inconsistencies produced by different rankings) are decided by minimizing the sum of the "strengths" of the inconsistencies (the rank difference between individuals whose dominance relationship is inconsistent). This method is implemented in MatMan (de Vries et al. 1993; Noldus Information Technology 2003; Section 2.9).

- Find the ranking that maximizes the sum, over dyads, of the win–loss difference multiplied by the difference in ranks (Crow 1990). This is an approximate maximum-likelihood solution when it is assumed that contests are independent and the probability of A defeating B is an increasing function of their difference in rank. Both assumptions may or may not be reasonable.

- Find the ranking that maximizes the likelihood (Section 2.8) of the results of the interactions under the condition of weak stochastic transitivity (WST) such that if A is likely to win over B, and B over C, then A is likely to win over C (McMahan & Morris 1984). This is theoretically a nice method, making a reasonable primary assumption about the nature of the dominance process. The two main drawbacks are that it also assumes independence of the interactions of a dyad, which may not be true in many cases, and that the computation required in maximizing the likelihood is not particularly straightforward (McMahan & Morris 1984). In addition, it may not produce a uniquely "best" ranking with several orderings having equal maximum likelihood.

Here are some dominance indices (there are others; Bayly et al. 2006):

- The number of animals dominated by individual I [v_I in Equation (16)] is simple, but it has the drawbacks that it cannot be calculated when there are unknown dominance relationships (e.g., a pair never observed interacting), and it does not use the interaction data if available.
- The proportion of the total encounters engaged in by an individual that it wins (Baker & Fox 1978) is also simple and easily calculated, but it might be very misleading if, for instance, individuals tend only to interact with those close to them in the hierarchy, leading to situations in which all except the most and least dominant animals win about half their contests. This problem is partly dealt with by Crook and Butterfield's (1970) index, which is the mean, over opponents, of the proportion of encounters won by an individual, and Zumpe and Michael's (1986) index, which is similar but combines aggressive and submissive behavior. These indices become problematic when encounters between dyads are few and/or very variable.
- David (1987) introduced methods of calculating dominance indices that deal with the situation in which interactions do not occur randomly across the hierarchy. In the version described by de Vries et al. (2006), David's score is given by

$$DS(I) = \sum_J D_{IJ} + \sum_J D_{IJ} \cdot \left(\sum_K D_{JK} \right) \sum_J D_{JI}$$

$$- \sum_J D_{JI} \cdot \left(\sum_K D_{KJ} \right) \tag{17}$$

Here, D_{IJ} is de Vries et al.'s (2006) dyadic dominance index [Equation (10)]. If $\sum D_{IJ}$ is thought of as I's net dominance and $\sum D_{JI}$ as I's net submissiveness. then $DS(I)$ can be interpreted as I's net dominance plus the sum of the net dominances of the animals that I dominated weighted by how much I dominated them, minus the corresponding values for net submissiveness. This is a generally good measure of interaction success (Gammell et al. 2003; de Vries et al. 2006).

· Clutton-Brock et al. (1979) introduced an index of fighting success that has some similarities to David's score. It only considers whether an individual ever won or lost to another individual, however, rather than the proportion of wins or losses, and divides the win measure by the lose measure, making it very nonlinear. Gammell et al. (2003) consider Clutton-Brock et al.'s (1979) index to be less satisfactory than David's score.

· The cardinal index of dominance rank of Boyd and Silk (1983) is a maximum-likelihood technique like the WST (McMahan & Morris 1984), but it makes more assumptions and can produce more-informative results. The basic assumption is that each individual I has a cardinal dominance index D_I and that the probability of winning in contests against J is a logistic function of $D_I - D_J$. The D_I's are estimated by maximum likelihood (Section 2.8), using an iterative procedure, and the method can give standard errors or confidence intervals for these, as well as likelihood ratio tests of a wide range of hypotheses about the dominance hierarchy (Boyd & Silk 1983). Weighing against these attractive features are the more rigorous assumptions of this technique, which include independence of interactions and a form of linearity in dominance, the iterative maximization of the likelihood, and the fact that the procedure often does not work when there are few entries below the diagonal (i.e., the hierarchy is nearly linear but also includes situations in which there are individuals who win or lose all their contests).

· The Elo rating method used by chess federations considers sequential data on dyadic encounters to update ratings of each individual in the population after each interaction (Albers & de Vries 2001). It makes sense to import a technique developed and accepted by a large number of quantitatively minded individuals engaged in dyadic contests. The Elo rating has a number of advantages over most other techniques used to characterize dominance hierarchies (Albers & de Vries 2001): It uses the information on individual contests, but it does not assume them to be independent; it uses sequential information; it can track (with some lag) temporal changes in dominance relationships (Fig. 5.11); it is operational on almost any data set, however sparse (i.e., lacking interactions between some pairs of individuals), for which sequential interaction information is available; it is unaffected by the pattern of frequency of interactions (e.g., individuals close or far apart in the hierarchy tending to interact more often); and it is quite easily implemented, containing no optimization routines or matrix algebra. I can see only two disadvantages: It needs sequential information that may not always be available, and the parameter k that determines the effect of each contest on an individual's Elo rating is arbitrary.

5.4.3: Dominance Hierarchies: An Example. To illustrate some of these measures, Table 5.8 summarizes the outcomes of pairwise agonistic interactions in a captive population of 10 female olive baboons (from Table I in McMahan & Morris 1984). There are eight dyads without pairwise data (indicated by 0s both above and below the diagonal in Table 5.8), and although there appears to be a dominance hierarchy, it is not perfectly linear. This is confirmed by a moderate value of $h' = 0.52$ that is marginally significant using de Vries' (1995) permutation test ($P = 0.06$). The expected value of h' under the null hypothesis of randomly distributed dominance relationships among 10 individuals is 0.27. The steepness (de Vries et al. 2006) of this hierarchy is 0.40, again a moderate value, but significantly larger than the expected steepness value of 0.21 ($P < 0.01$), and so this might be characterized as a "somewhat despotic" society.

Table 5.9 compares dominance ranks and dominance indices for the baboon data [de Vries (1998), de Vries and Appleby (2000), Gammell et al. (2003), and Bayly et al. (2006) also provide comparisons of the performances of some of these and other measures]. Of the dominance indices in the foregoing list, the number of animals dominated is not applicable because of the lack of information for some dyads, the car-

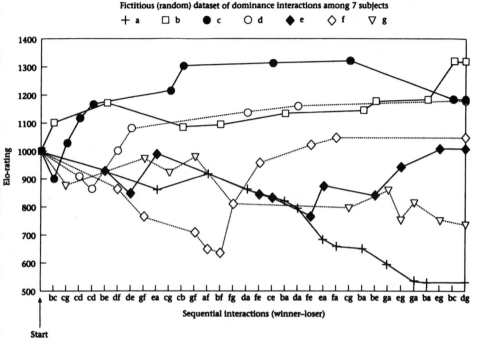

FIGURE 5.11 Example of using the Elo rating to describe a dominance hierarchy from 33 sequentially ordered dyadic interactions among seven individuals in a fictitious data set. (From Albers & de Vries 2001, fig. 1.)

dinal dominance rank of Boyd and Silk (1983) cannot be calculated because there is an individual (#907) that won all her contests, and Elo ratings cannot be calculated because sequential information is not available.

The results of the different measures are consistent with regard to the ranking of the most (#907) and least (#902) dominant individuals, but in other respects there is considerable discordance (Table 5.9). The different methods emphasize or deemphasize different aspects of the data. For instance, the ranking of #915 fourth or fifth using Crow's scheme (as opposed to second or third on all others) is driven by a single loss to the usually lowly ranked #897. Of the dominance rankings, only the I&SI method provides a unique ranking, which agrees with one other chosen ranking of the other ranking methods, except Crow's. Note that the I&SI method does not necessarily produce a unique ranking (de Vries & Appleby 2000). Ranks derived from the three dominance indices differ from each other and from the directly derived dominance ranks.

Table 5.8 Outcomes of Pairwise Agonistic Interactions in a Captive Population of 10 Female Olive baboons (*Papio cynocephalus*)

		Loser									
		#907	#915	#912	#910	#917	#898	#897	#911	#904	#902
	#907	–	2	0	5	2	2	1	0	2	0
	#915	0	–	2	2	1	0	3	2	1	1
	#912	0	1	–	1	1	3	1	1	4	0
Winner	#910	0	0	0	–	1	1	1	0	1	0
	#917	0	0	0	0	–	7	1	4	2	3
	#898	0	0	0	0	0	–	2	3	6	10
	#897	0	1	1	0	2	0	–	0	0	2
	#911	0	0	0	1	0	0	0	–	1	1
	#904	0	0	0	1	0	0	0	0	–	1
	#902	0	0	0	0	0	0	0	0	0	–

From McMahan and Morris (1984).

Table 5.9 Dominance Ranks and Dominance Indices for Baboon Data from Table 5.8

	Dominance rankings					Dominance indices		
	I	Brown	I&SI	Crow	WST	PCW	David	CBI
907	1	1	1	1	1	1	19.2	33
#915	2	2	2	4,5	2	0.75	12.2	2.75
#912	3	3	3	3	3	0.8	12.5	3.08
#910	4	4	4	7	4	0.29	−1.3	0.91
#917	5,6,7	5	5	2	5	0.71	2.9	0.86
#898	5,6,7	6	6	4,5	6	0.62	−0.7	0.82
#897	5,6,7,8,9	7,8,9	8	6	7,8	0.4	−2.9	0.92
#911	7,8	7,8	7	8	7,8,9	0.23	−7.6	0.53
#904	8,9	8,9	9	9	8,9	0.11	−14.6	0.23
#902	10	10	10	10	10	0	−19.9	0.03

When two or more dominance ranks were chosen by a criterion, all the ranks of an individual are shown. The dominance ranks are as follows: I (Slater 1961); Brown (Brown 1975, p. 86); I&SI (de Vries 1998); Crow (Crow 1990); WST (weak stochastic transitivity), (McMahan & Morris 1984). The dominance indices are the proportion of contests won (PCW), David's (1987) score as modified by de Vries et al. (2006), and Clutton-Brock et al.'s (1979) index (CBI).

BOX 5.4 *Analyzing Dominance Hierarchies: Recommendations*

Dominance data are usefully summarized by the sociometric interaction matrices illustrated in Tables 5.7 and 5.8. The linearity of the hierarchy should usually be assessed by Landau's (1951) *h* or, when there is no information for some dyads, de Vries' (1995) *h'*. Except in cases of very obvious hierarchies or the lack of them, testing the null hypothesis of random dyadic dominance against the alternative of a hierarchy using the methods of Appleby (1983) or de Vries (1995) gives an important perspective on the data. Often, it will also be useful to calculate de Vries et al.'s (2006) steepness as a measure of how egalitarian or despotic a society is. If h (or h') is reasonably high and larger than would be expected from random dominance relationships, then obtaining a dominance index and/or dominance ranking for the individuals is usually desirable.

The proliferation of ranking methods, some of which need a fair amount of quite sophisticated computation, will be off-putting to some social analysts. It is worth remembering that if the hierarchy is nearly linear (with h or $h' > \sim 0.9$), then the ranking should be fairly obvious by any method, whereas if it is less linear, then the discordance among ranking methods usually reflects a real lack of predictability in the outcomes of interactions in the population. I follow Gammell et al. (2003) in recommending David's score as providing a dominance index, and thereby dominance ranking, that has generally desirable properties and can be quite easily calculated (even by hand). The version of David's score introduced by de Vries et al. (2006) [Equation (17)] is probably best because it considers the amount of data available for each dyad. David's score is also needed for calculating steepness. The I&SI (de Vries 1998) technique seems to be in some ways the best of the straight ranking techniques (de Vries & Appleby 2000), and it is calculated by MatMan (de Vries et al. 1993). In particular circumstances, other techniques may be useful. For instance, if the fairly restrictive requirements of Boyd and Silk (1983) are met, then their cardinal index has many useful properties, and if sequential data are available, then the chess-players' Elo ranking (Albers & de Vries 2001) could be very informative.

Martin and Bateson (2007, p. 134) urge caution in interpreting dominance hierarchies. Dominance relationships can differ, depending on the interaction measure used; they can change temporally (e.g., Table 5.7) and even spatially, with, for instance, the results of dominance interactions between two territorial animals being dependent on the relative distances to the centers of their territories.

5.5 Adding Time: Lagged Association Rates

Temporal patterning is one of the elements of Hinde's (1976) characterization of social relationships. How associations, interactions, and relationships change over time are key components of the social life of an animal (Section 4.6) and the nature of the society that it inhabits. Traditional ordinations, network analyses, and dominance hierarchy analyses indicate temporal patterning coarsely at best. For instance, ordinations, cluster analyses, network analyses, and dominance hierarchy analyses can be made in different time periods and compared (Sections 5.2–5.4). Using the Elo rating to describe a dominance hierarchy can provide a temporal axis to an aspect of sociality (Fig. 5.11), but this is an imprecise, lagged description. The temporal scales of sociality are not addressed.

Lagged association rate analyses, which are based on the methods of Underwood (1981) and Myers (1983), were designed to address the scales of temporal patterning in social relationships (Whitehead 1995). Introduced in Section 4.6 as a method for describing the temporal patterning of a dyadic relationship, their principal utility has been at the societal level. Partly this is because there are usually insufficient data to use these methods effectively for a single dyad, and partly because they integrate well over populations. In addition, by fitting models to lagged association rates, we can potentially both uncover structural aspects of a social organization and estimate the parameters of that structure. However, it is important to be aware that different social structures can give rise to similarly patterned lagged association rates.

In the following subsections, I show how to calculate lagged association rates, discuss some variants and options, and then consider model fitting.

5.5.1: The Lagged Association Rate. The dyadic lagged association rate of lag τ is simply the probability of association τ time units after a previous association and is estimated by the ratio of the number of associations X has with Y τ time units apart divided by the number of pairs of identifications of X τ time units apart (Section 4.6). This is generalized to the population lagged association rate $g(\tau)$, the probability of association τ time units after a previous association averaged over all associations. It is estimated by taking the ratio of the number of observed dyadic associations that were repeated τ time units apart over the potential number of such dyadic associations, given when each individual was identified. Technically, it is given by (Whitehead 1995)

$$g(\tau) = \frac{\displaystyle\sum_{j,k|(t_k-t_j)=\tau} \sum_{X} \sum_{Y\neq X} a_j(\mathbf{X},\mathbf{Y}) \cdot a_k(\mathbf{X},\mathbf{Y})}{\displaystyle\sum_{j,k|(t_k-t_j)=\tau} \sum_{X} \sum_{Y\neq X} a_j(\mathbf{X},\mathbf{Y}) \cdot a_k(\mathbf{X},\mathbf{X})} \tag{18}$$

where $a_j(X, Y) = 1$ if X and Y were recorded as associated in time period j, $a_j(X, Y) = 0$ if they were not associated or if either was not identified during the sampling period, $a_k(X, X) = 1$ if X was identified in period k, and $a_k(X, X) = 0$ if X was not identified in period k. Summing over each pair of sampling periods τ time units apart, j and k, and over all individuals (X) identified in both periods, we obtain the numerator as the total number of other individuals identified in the same group as X in both periods, whereas the denominator is the total number of other individuals identified in the same group as X in the first period.

The lagged association rate $g(\tau)$ is often plotted against lag τ to describe how relationships change with time (e.g., Fig. 5.14). If the range of τ considered is wide (say, more than an order of magnitude), then it is usually clearer to log the x-axes, the time lag. The y-axis may usefully be logged in some circumstances (such as when there are two or more processes occurring at very different scales).

Usually, sample sizes are too small for some values of τ to give an estimate of $g(\tau)$ with acceptable precision. Thus, it makes sense to lump the data for a range of τ or employ a moving average (as in SOCPROG), widening the range of τ being considered so that the denominator of Equation (18) is above some minimum. The desirable minimum depends on the situation, and so the analyst should try several values to find a good compromise between the precision of $g(\tau)$ and the precision of τ. A great deal of lumping will produce a smooth curve of lagged association rates but be imprecise for any value of τ; too little lumping will give a curve with many spurious peaks and troughs (Fig. 5.12).

Like any other social measure, lagged association rates have little value without some measure of precision. There are no analytical estimates of confidence intervals for $g(\tau)$, and I have not found a valid nonparametric bootstrap procedure. Of the methods that I have experimented with, the parametric bootstrap (Section 2.3) performs best, but it requires a mathematical model of the social structure and is cumbersome to implement. More usable, but rather approximate, is the temporal jackknife (Section 2.3) in which different sets of sampling periods are omitted in turn (Whitehead 2007). This is the method implemented by SOCPROG (Fig. 5.16).

5.5.2: Standardized Lagged Association Rate. In many cases, although our records of associations are accurate, a zero in the association data only means that a dyad was not observed to associate in the sampling period, not that they did not associate. In addition, the rate of identifying

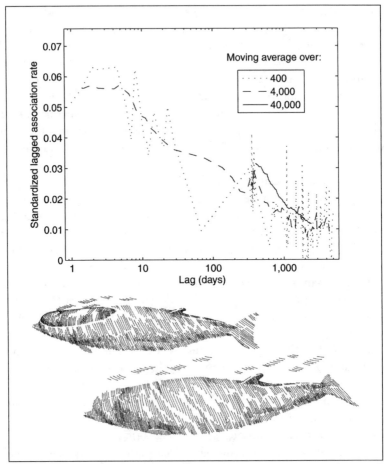

FIGURE 5.12 Standardized lagged association rates for 243 northern bottlenose whales (*Hyperoodon ampullatus*) with various degrees of lumping (association defined as observed within 1 hour, sampling periods of 1 day). With a moving average of 400 associations, the plot is too spiky; with 40,000 associations, much information is lost. The line with a moving average of 4,000 associations is close to optimal. (Illustration copyright Emese Kazár.)

individuals and thus associates may vary among sampling periods. In such cases, the probability of two individuals being associated after some lag is not estimated by Equation (18). Instead, in these situations, I suggest estimating a related quantity, the standardized lagged association rate (Whitehead 1995). It is the probability that, supposing Y is an associate of X, if a randomly chosen associate of X is identified after a lag of τ, then it is Y. The standardized lagged association rate is estimated from

$$g'(\tau) = \frac{\displaystyle\sum_{j,k|(t_k-t_j)=\tau} \sum_{X} \sum_{Y \neq X} a_j(\mathbf{X}, \mathbf{Y}) \cdot a_k(\mathbf{X}, \mathbf{Y})}{\displaystyle\sum_{j,k|(t_k-t_j)=\tau} \sum_{X} \left(\sum_{Y \neq X} a_j(\mathbf{X}, \mathbf{Y}) \right) \cdot \left(\sum_{Y \neq X} a_k(\mathbf{X}, \mathbf{Y}) \right)} \qquad (19)$$

where the notation is as for Equation (18).

Although the lagged association rate should be 1.0 at time lags so small that no associations have changed, the standardized lagged association rate at very small lags will be the inverse of the mean number of associates of an individual during a sampling period, its gregariousness or typical group size minus one. Standardized lagged association rates are plotted against lag, as with lagged association rates, and precision can be estimated using the temporal jackknife method.

5.5.3: Null Association Rate. To interpret lagged association rates and standardized lagged association rates, it helps to consider what values they would have if animals associated randomly. I call these the null association rates and standardized null association rates, respectively. The null association rate is the ratio of the gregariousness of the population (i.e., mean number of associates of an individual in any sampling period; Section 4.3) to the number of identified individuals minus one. SOCPROG calculates the null association rate using the association data available for each time lag, so it varies a little with τ. The standardized null association rate is the inverse of the number of identified individuals minus one, and does not change with time lag.

5.5.4: Intermediate Association Rate. I proposed the intermediate association rate (Whitehead 1995) as a statistic that would help to distinguish between social systems that give the same pattern of lagged association rates (for some examples of such indeterminacy, see the description of Fig. 5.13). The intermediate association rate is calculated in a similar way to the lagged association rate, but, between any two individuals, only associations [the numerator in Equation (18)] and potential associations [the denominator in Equation (18)] between the first and last recorded association, and only intervals including either the first or last association, are considered (Whitehead 1995). For intermediate association rates, the time lag is the minimum of the time between the sampling period and the first or last association. Intermediate association rates approximate 1.0 if associations with long lags are between members of permanent units that do not disassociate between observed associations. If long-term reassociations often follow periods of separation, then the intermediate association rate's relationship with time lag may be similar to that of the lagged association rate. The standardized intermediate association rate is calculated similarly to the lagged association rate but includes the number of associates of each individual at both the beginning and end of the time lag in the denominator, as in Equation (19).

One needs a fair amount of data to compute meaningful intermediate association rates or standardized intermediate association rates, but if they are available, they can help to interpret lagged association rates. The expected values of intermediate association rates can depend on the amount of data collected, however, which limits their usefulness. For instance, consider a society in which dyads associate continuously for a while and then break up but may reassociate after a long time period. Then, if the sampling rate is such that the data usually contain no more than one period of association per dyad, the intermediate association rate will stay around 1.0. In contrast, if the data usually include several periods of association per dyad, then the intermediate rate will fall with lag, showing a similar pattern to the lagged association rate. For these reasons, I now believe that the value of the intermediate association rate is rather less than when I originally proposed it (Whitehead 1995), and I do not include it in the examples that follow. A revised version of the intermediate association rate that is less biased by sampling intensity could probably be developed but has not yet been.

5.5.5: Lagged Identification Rate. Another related measure that can help in the interpretation of the lagged association rates is the nonsocial lagged identification rate (Whitehead 2001). The lagged identification rate informs us about movements into and out of a study area. It estimates the probability that an individual in the study area at any time is the same as a randomly chosen individual from the study area τ time units later, and thus is the probability of remaining in the study area divided by the population size in the study area. It is estimated from (Whitehead 2001)

$$R(\tau) = \frac{\displaystyle\sum_{j,k|(t_k - t_j)=\tau} m_{jk}}{\displaystyle\sum_{j,k|(t_k - t_j)=\tau} n_j \cdot n_k} \tag{20}$$

where n_j is the number of individuals identified in sampling period j and m_{jk} is the number of individuals identified in both periods j and k. There is no lagged-identification-rate analog to the standardization of lagged association rates, because the lagged identification rate is, in this sense (having identifications rather than associations in the denominator), already standardized.

If the population is closed, with no birth, death, immigration, or emigration, then the lagged identification rate should be constant at the inverse of the population size minus one. A fall in lagged identification rate over lags of about T time units indicates animals leaving the population through emigration or mortality (or possibly misidentifications) at overall rates of roughly $1/T$ per time unit [more precise estimates of leaving rates can come from fitting models; see later discussion and Whitehead (2001)]. If, following a fall, the lagged identification rate starts to level again, this may indicate emigration from, and then reimmigration into, the study area or a mixed population of residents and transients. Confidence intervals for lagged identification rates can perhaps best be estimated using the bootstrap technique (Whitehead 2007) in which individuals are sampled with replacement to obtain bootstrap replicates (Section 2.3).

Lagged identification rates help to place lagged association rates in perspective because, if one individual has left the population, then it cannot associate with anyone still in it, and if two individuals have left the population, then their association pattern is unknown.

5.5.6: Interpreting Lagged Association Rates. To illustrate the interpretation of plots of lagged association rates and related measures (null association rates and lagged identification rates), Fig. 5.13 shows six scenarios:

1. Figure 5.13A: "closed, non-interacting units." Here, the lagged association rate is constant at 1.0 at all lags, indicating that all associations persist throughout the study, and so there are completely closed, noninteracting units. There is no information in the plot about the sizes of the units.

2. Figure 5.13B: "casual acquaintances." The lagged association rate falls to the null association rate over periods of about 10 time units, suggesting casual acquaintances who associate for about this period before breaking up. They may reassociate later.

3. Figure 5.13C: "constant companions plus casual acquaintances." In this scenario, the lagged association rate falls but stabilizes above the null association rate. Several social systems could produce this pattern, including animals forming permanent social units that themselves associate for periods

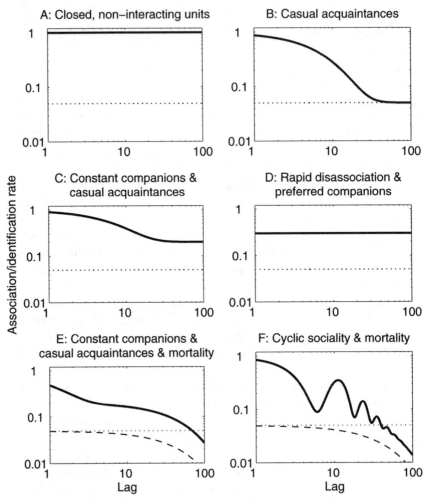

FIGURE 5.13 Possible profiles of lagged association rates (*solid line*), null association rates (*dotted line*), and lagged identification rates (*dashed line*) indicating different social structures.

of about 10 time periods; casual but preferred acquaintances in which associations break up after about 10 time periods, but individuals are more likely to associate with those with whom they have previously associated than with members of the population at random; and a situation in which units have a permanent core membership but there are also "floaters" who move between units.

4. Figure 5.13D: "rapid disassociation plus preferred companions." Here, the lagged association rate is constant over the

time lags examined but less than 1.0 and greater than the null association rate. In this situation, many associations are broken over less than one time period, but there is little change over the range of lags considered. This is consistent with social systems in which associations are ephemeral, but there are preferred associations, or a population that contains no social preference but is geographically structured so that individuals are more likely to associate with some members of the population than others.

5. Figure 5.13E: "constant companions plus casual acquaintances plus mortality." Here, the lagged association rate, when plotted on a log scale, shows two falls, first over periods of one to two time periods, and then over greater than about 40 time periods. The first drop, at small lags, could be due to the kind of process discussed for Fig. 5.13C, such as permanent but associating social units, casual but preferred acquaintances, or permanent units plus floaters. The second drop over the longer lags coincides with a similar drop in lagged identification rates, indicating that this is a demographic feature: Individuals are dying or leaving the identified population over such lags. If the lagged identification rate did not fall over such lags, then the two-drop lagged association rate would indicate two types of social disassociation over different time scales, such as might be caused by semipermanent units that associate with each other and retain their membership over substantial time periods but do eventually break up.

6. Figure 5.13F: "cyclic sociality plus mortality." This shows a more complex pattern, including cyclic association and disassociation over periods of about 12 time units, plus mortality or emigration. To clearly detect such a pattern would require considerable data, and I have never seen such a situation in practice.

If the lagged association rates and null association rates in Fig. 5.13 were replaced by standardized lagged association rates and standardized null association rates, we would interpret the patterns similarly in most cases. Because standardized lagged association rates are not fixed at 1.0 with zero lag, however, the social systems that produced the patterns in Fig. 5.13A and Fig. 5.13D are not distinguishable based on standardized rates.

Figure 5.14 shows a variety of patterns of lagged association and standardized lagged association rates from published papers. The majority of these are of cetaceans because lagged association rates have been primarily used on these species. All show a fall with time lag, but the drops occur at about 12 hours for eland and pilot whales and 100 days for the disk-winged bats and bottlenose dolphins, whereas with the killer whales and spinner dolphins, the fall is more gradual and seems to occur over a wide range of time lags. In the case of the spinner dolphins, the drop could be due to mortality (Karczmarski et al. 2005). The lagged association rate or standardized lagged association rate is well above the null association rate (not shown in Fig. 5.14 for clarity), indicating nonrandom associations in all these cases, except for the bottlenose dolphins at the longest lags of about 6 years.

5.5.7: Fitting Models to Lagged Association Rate. Lagged association rate analyses can give more than a display of the temporal patterning of a social structure. We can fit models to lagged association rate data, and thus have some quantitative basis for accepting or rejecting the presence of certain social elements. Model fitting also allows us to make quantitative estimates of social parameters, so the interpretation of Fig. 5.13C is changed from "associations break up after about 10 time periods" to "the mean duration of associations is estimated to be 8.3 time units (standard error 2.1)". Model fitting is useful, but the results should not be overinterpreted. The model chosen describes the lagged association rates, not social structure per se, and, as emphasized in the previous subsection, different social structures can give qualitatively and quantitatively similar patterns of lagged association rates.

There are some technical challenges with fitting models to lagged association rates that are beyond the scope of this book. They revolve around the issue that, even with a fairly simple model of social structure, if there are more than just a few sampling periods, then the number of possible identification histories of an individual becomes too large for even modern computers to handle. Ways around this are discussed by Whitehead (2001) in the case of fitting models to lagged identification rate data, but these also apply to fitting models to lagged association rates, and are implemented by SOCPROG. They basically amount to collating information for each pair of sampling periods, assuming that the number of repeat associations for each pair of periods [the numerator of Equation (18) inside the first summation sign] is drawn from a binomial distribution. The binomial parameters are the number of possible periods [denominator of Equation (18) inside the first summation

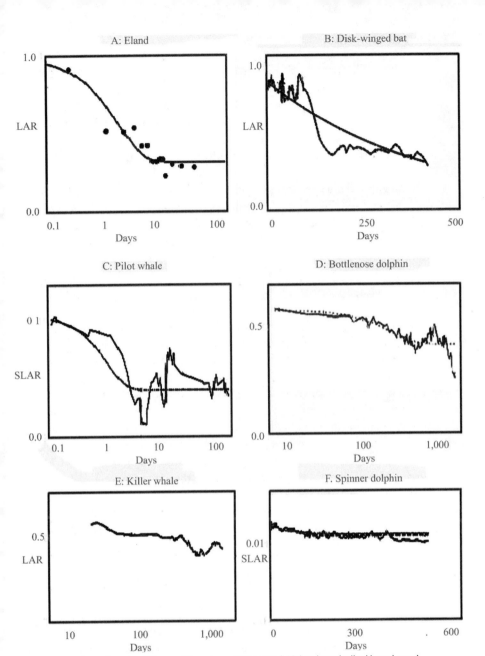

FIGURE 5.14 Examples of the use of lagged association rates (LARs) and standardized lagged association rates (SLARs). **A:** Eland (*Taurotragus* sp.; data from Underwood 1981). **B:** Disk-winged bats (*Thyroptera tricolor*; data from Vonhof et al. 2004, fig. 5). **C:** Pilot whales (*Globicephala melas*; data from Ottensmeyer & Whitehead 2003, fig. 3). **D:** Bottlenose dolphins (*Tursiops* spp., data from Lusseau et al. 2003, fig. 6a). **E:** "Transient" killer whales (*Orcinus orca*, data from Baird & Whitehead 2000). **F:** Spinner dolphins (*Stenella longirostris*, data from Karczmarski et al. 2005, fig. 4, upper panel).

sign] as sample size and the expected lagged association rate $g(\tau)$ as the probability of a repeat association. Then models can be fit by choosing parameters that maximize the likelihood of these data (Section 2.8).

The principal problem with this procedure is that the data as used are not independent, so that an assumption of the method of likelihood is violated. I have shown analytically and using simulation (Whitehead 2001), however, that the parameter estimates produced by maximum likelihood are approximately unbiased. A version of the likelihood method, quasilikelihood, that attempts to compensate for nonindependence of count data (Burnham & Anderson 2002, pp. 67–70) can be used to produce estimated confidence intervals, but the temporal jackknife (Section 2.3), which removes different sets of sampling periods in turn and can be used to estimate confidence for the lagged association rates (see earlier discussion), is also a good option. The quasilikelihood version of AIC, QAIC (Section 2.8), can be used to select among different models of lagged association rates (Whitehead 2007).

Which models should be fitted to lagged association rates? Those that I have found useful are of the exponential family in which lagged association rates are built up from processes whose effects (forming an association or breaking it) are equally likely to occur in any time interval. These models can cover a range of possible scenarios, as I show, but nonexponential models could be useful in special cases. For instance, with a lagged association rate pattern such as that shown in Fig. 5.13F, cyclical patterns could be generated by a model including trigonometric functions of the lag, such as $\cos(\tau)$. Here are the exponential models for lagged association rates that SOCPROG fits (there is also an option to fit custom models). The models are in pairs—the same basic model with and without rapid disassociation over periods of less than one sampling period (represented by the first two columns in Fig. 5.15):

- LAR1: $g = 1$. Here, all associations persist through out the study. This is the "closed, noninteracting units" pattern of Fig. 5.13A.
- LAR2: $g = a$. Here, some associations decay rapidly within one sampling period, and then the lagged association rate is stable. If $a \approx m/(n-1)$, the null association rate (where m is the mean gregariousness of the population), then there is no sign of preferred or prolonged association in the population. If $a > m/(n-1)$, which is the pattern shown in Fig. 5.13D, then the model is that to be expected when individuals associate ephemerally but nonrandomly, either because

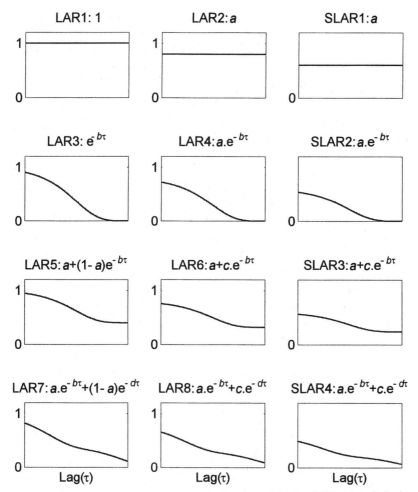

FIGURE 5.15 Exponential models that can be fitted to lagged association rates (LARs) and standardized lagged association rates (SLARs).

of dyadic preference or structural features such as ranging patterns.

· LAR3: $g = e^{-b\tau}$. This is a model of straight exponential decay and is easiest to imagine in a very large population in which animals associate for about $1/b$ time units and then never again. However, it is possible to envisage other scenarios producing lagged association rates that fit this pattern, for instance, a smaller population where animals avoid previous associates or a population with a very high mortality or emigration rate of about b per sampling period. In the latter case, the lagged identification rate should also fall over time lags of about $1/b$.

- LAR4: $g = a \cdot e^{-b\tau}$. This combines LAR2, indicating a proportion a of associations that disassociate within a single sampling period, with LAR3, so that associations that do persist beyond one sampling unit eventually fall to zero either because of random association in a large population or active avoidance of previous associates or departure from the population.

- LAR5: $g = a + (1 - a) \cdot e^{-b\tau}$. Here, association rates initially fall exponentially but then level off. This could indicate the scenario of casual acquaintances within a closed population of Fig. 5.13B if the leveling out is at the null association rate $[a \approx m/(n - 1)]$. If the long-term lagged association rate is above the null association rate $[a > m/(n - 1)]$, then this indicates the "constant companions plus casual acquaintances" scenario of Fig. 5.13C. This can be produced by social systems such as permanent social units (of approximate typical unit size $u = 1 + m \cdot a$ if the population is fairly large) that associate temporarily (for periods of about $1/b$ sampling periods), casual, but preferred, associations lasting about $1/b$ sampling periods, and a situation in which some individuals form permanent units, whereas others are "floaters" who move between units.

- LAR6: $g = a + c \cdot e^{-b\tau}$. This combines models LAR2, with rapid disassociation within one sampling period, and LAR5, in which association rates fall and then level off. The possibilities listed under LAR5 apply here, with rapid disassociation added to them.

- LAR7: $g = a \cdot e^{-b\tau} + (1 - a) \cdot e^{-d\tau}$. Here, there are two levels of disassociation at time scales of $1/b$ and $1/d$, respectively, as shown in Fig. 5.13E. The shorter will probably be a social disaffiliation, of the types discussed under LAR5, but the more permanent relationships, either within units or preferential association patterns, eventually decay for reasons such as some movement between permanent units, shifts in preferred companionship, mortality, emigration, or combinations of these.

- LAR8: $g = a \cdot e^{-b\tau} + c \cdot e^{-d\tau}$. This combines the rapid disassociation of LAR2 with the two levels of disassociation of LAR7.

Similar models can be fit to standardized lagged association rates, although there are some differences, particularly in that rapid disassociation cannot be directly incorporated because this is confounded with gregariousness. Thus, each model of standardized lagged association rates is analogous to two models of unstandardized lagged association rates, with and without rapid disassociation (given along the rows of Fig. 5.15). Here are the standard models that SOCPROG can fit to standardized lagged association rates:

- SLAR1: $g' = a$. Here, there is no change in association rate with lags of one sampling period or more. This could represent the "closed, noninteracting units" pattern of LAR1, in which case a is the inverse of the gregariousness, or if there is rapid disassociation, as in LAR2, a is less than the inverse of the gregariousness. In the extreme, with no associations persisting over one sampling period in a closed population of size n, then $a = 1/(n - 1)$.

- SLAR2: $g' = a \cdot e^{-b\tau}$. This indicates casual acquaintances in a large population, possibly including rapid disassociation, as in LAR3 and LAR4. The duration of associations is of the order of $1/b$, and if there are no rapid disassociations, a is the inverse of the gregariousness.

- SLAR3: $g' = a + c \cdot e^{-b\tau}$. Here, association rates fall with time lag and then level off—the patterns in LAR5 and LAR6—and the explanations given for these models also hold here. The duration of associations is of the order of $1/b$, and if there is no rapid disassociation, then the gregariousness is $1/(a + c)$, and, in the case of permanent units that temporarily group, the typical unit size is about $1 + a/(a + c)^2$.

- SLAR4: $g' = a \cdot e^{-b\tau} + c \cdot e^{-d\tau}$. Here, there are two levels of disassociation, perhaps the fission/fusion of nearly permanent social units into and out of groups on the short time scale and transfers between units on the longer one. Other scenarios producing this pattern were suggested for LAR7 and LAR8.

Some technical issues may come into play when fitting models of lagged association rates and standardized lagged association rates. The model fitting is by iterative convergence to the original association data (not estimated lagged association rates). One may have to try different

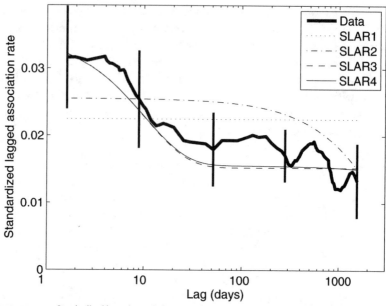

FIGURE 5.16 Standardized lagged association rates plotted against time lag for 1,300 female and immature sperm whales (*Physeter macrocephalus*) studied in the South Pacific between 1985 and 1995. Vertical
bars indicate approximate standard errors calculated using the temporal jackknife method (omitting each
year's data in turn). A moving average of 100,000 associations was used to smooth the curve. The best-
fitting models of four members of the exponential family are shown (see table 5.10 for formulas). These
models were fit using maximum likelihood to the original data, not to the calculated standardized lagged
association rates.

start positions for the parameters *a*, *b*, *c*, and *d* to get a convergence and
a satisfactory model fit. With large data sets over long time intervals
(e.g., 100 individuals frequently observed over 10 years), using a short
sampling period (say, 15 minutes) may produce too much data for
one's computer to handle. In these cases, one should place a cap on the
maximum time scale being investigated (e.g., 12 hours; SOCPROG has
an option for this), and then examine lags of greater than 1 day using
a larger sampling period, maybe 1 day. The two lagged association rate
plots can then be joined to show how associations change over scales
from 15 minutes to 10 years. One should use a logged x-axis in this case.

As an example, Fig. 5.16 shows standardized lagged association rates
for female and immature sperm whales, with the four exponential models for standardized lagged association rates fitted. The two more complex models show reasonable fit to the data, and this is confirmed by the
QAIC analysis shown in Table 5.10. The "constant companions plus
casual acquaintances model," SLAR3, has the lowest QAIC and so fits
best. Adding a second level of disassociation or mortality, SLAR4, gives

Table 5.10 Exponential Models (τ Is Time Lag in Days) Fit to Data on Standardized Lagged Association Rates (g') of Sperm Whales as Shown in Figure 5.16

Model	Best fit	QAIC	ΔQAIC	
SLAR1	$g' = 0.022$	5,686.93	60.75	No support
SLAR2	$g' = 0.026e^{-0.00034\tau}$	5,660.67	34.49	No support
SLAR3	$g' = 0.015 + 0.020e^{-0.095\tau}$	5,626.18	0	Best
SLAR4	$g' = 0.016e^{-0.000022t\tau} + 0.019e^{-0.097\tau}$	5,628.12	1.94	Some support

The lowest quasilikelihood Akaike Information criterion (QAIC) indicates the best-fitting model, and ΔQAIC (difference between QAIC and that of the best model) indicates the degree of support for the other models.

almost the same curve (Fig. 5.16) and a QAIC higher by 1.94, indicating some support. There is virtually no support for the simpler models, SLAR1 and SLAR2 ΔQAIC > 30; see Section 2.8).

These models suggest that there is important disassociation over scales of about 10 days ($1/b$), but that the stabilization at lags greater than this ($g' \approx 0.015$; Fig. 5.16) is well above the standardized null association rate [$0.0008 = 1/(1,300 - 1)$]. Sociograms, cluster analyses, intermediate association rates, and simple inspection of the data show that this is because of temporary merging of nearly permanent social units (e.g., Whitehead et al. 1991), and so, using the formulas given with the description of SLAR3, we obtain an estimate of the typical group size of 29.6 animals, and the typical unit size 13.4 animals. If SLAR3 is reparameterized as $a = (u - 1)/m^2$ and $c = (m + 1 - u)/m^2$, where m is the gregariousness (so $m + 1$ is the typical group size) and u is the typical unit size, then the jackknife method gives standard errors for the measures of social structure: $m + 1 = 29.6$ (SE 6.6), $u = 13.4$ (SE 4.2) and $b = 0.095$/day (SE 0.046). Furthermore, there are indications of a fall in the standardized lagged association rate at high lags in Fig. 5.16, and, if we accept SLAR4, this is estimated to be at a rate of 0.000022/day (SE 0.00018499) = 0.008/year (SE 0.0675). This is not an unreasonable, if a very inaccurate, estimate of mortality for sperm whales.

BOX 5.5 *Lagged Association Rates: Recommendations and Extensions*

The display of lagged association rates, and related methods described in Section 5.5, constitute the major currently available analytical tool that allows the social analyst to address the issue of temporal scale directly. Because temporal scale is a vital element of social structure, this gives these techniques a potentially important role in the social analysis of any species. For species with large ranges and large populations, however, like cetaceans, their relative significance is increased. This is because, unlike ordinations and network analyses, lagged association rates naturally integrate over large, sparse data sets. When many individuals are just identified once or twice, association indices or interaction rates are highly imprecise (Section 3.11), and any display produced using them may be extremely misleading. Usually, then, we restrict attention to just those individuals with reasonably large sample sizes (Box 5.1), but is this biasing our sample or giving a false impression? In contrast, lagged association rates use all of the data (do not restrict to individuals seen more than x times when carrying out lagged association rate analyses; Box 5.1). If there are much data in total, even though there are only little data on any individual and even less on any dyad, we may get a useful picture of temporal aspects of relationships and be able to estimate social measures for the population. As an example, the 1,300 sperm whales used in the analysis of Fig. 5.16 were identified on an average of 1.7 days each, yet an informative model of social structure was produced (Section 5.5).

There are, however, limitations and difficulties in the use of lagged association rates. These analyses do need relatively large amounts of data to be useful. With just a few identifications of a few individuals, lagged association rates will be uninformative. Several technical issues need consideration:

- The definitions of association and sampling period. These definitions are crucial to the analysis. The shorter the sampling period, the finer is the scale of the temporal analysis. However, a very short sampling period may give too much data (too many pairs of associations) over longer time scales. The way around this is only to consider short lags with the fine sampling period and meld the results to those from a longer-scale analysis with a coarser sampling period.
- Should one use lagged association rates or standardized lagged association rates? If the lack of an association in a sampling period can be recorded accurately, then lagged association rates can be used. Their meaning is more intuitive and they are more

informative. For instance, a greater range of exponential models can be fitted (Fig. 5.15). If the social
analyst cannot be sure that a dyad did not associate
in a sampling period even though one or both members were identified, then she should use the standardized lagged association rates.

- Pooling. Over what time lags should data be pooled?
  With little pooling, the lagged association rates may
  be very "spiky," with the spikes having no social relevance (Fig. 5.12). With too much pooling, the temporal resolution is lost. My practice is to increase the
  pooling (number of associations in the moving average in SOCPROG) until a fairly smooth and believ-
  able curve emerges (Fig. 5.12). I have found no a priori rule of thumb as to how large the pooling should
  be.

- Standard errors. Standard errors in lagged association rates can be obtained using the temporal
  jackknife (as in Fig. 5.16), but we need to choose
  temporal periods on which to jackknife that can be
  considered reasonably independent. This method
  produces only rough standard errors. The standard
  errors of parameter estimates of fitted models can
  also be estimated using the temporal jackknife or by
  quasilikelihood methods. These standard errors are
  approximate and are not properly justified theoretically (Whitehead 2007).

- Which models to fit? Burnham and Anderson (2002,
  p. 19) warn against trying to fit too many models.
  This should not be too great a problem if potential social structures for the population are used
  as the basis for choosing potential models. The
  exponential-family models of lagged association
  rates and standardized lagged association rates
  considered by SOCPROG and described in Section 5.5 (see also Fig. 5.15) are appropriate in many
  cases.

- Model selection. Although using the QAIC to select
  models of lagged association rates and standardized lagged association rates can be useful in many

circumstances, it is not fully theoretically justified
and can be misleading (Whitehead 2007). Thus,
QAICs should be considered a rough guide to the
value of the different models.

The most important caution when using lagged association rates
and standardized lagged association rates is that these measures
describe aspects of the social structure, they do not prescribe them.
As emphasized in Section 5.5, several quite different social structures can produce the same patterns and fitted models of lagged
association rates. Thus, lagged association rates may be important
tools in examining social structures, but they do not tell the whole
story and should be used in conjunction with other methods such
as ordinations (Section 5.2) and network analyses (Section 5.3).

Lagged association rate analyses can be extended to look at
between-class associations, as described in Section 4.11. Thus, we
can, for example, examine the changes over time in associations
between males and females (e.g., Fig. 4.6). Another extension,
which has not yet been developed but should be feasible, is to
produce a version of the lagged association rate for interaction,
rather than association, data. Then we would be asking questions
such as, How do dyadic grooming rates change with time lag?

A potentially exciting development is in the work of Palla et al.
(2007) who explicitly incorporates temporal change and the evolution of social groups into network analysis. Although their
examples are for binary networks, they indicate that the method
is usable on weighted networks.

5.6 Multivariate Methods

The methods considered so far in this chapter have used data from one
association measure or interaction type. Our depiction of a relationship
is more powerful if we have more than one interaction or association
measure (Section 4.7), and, consequently, models of social structure are
more powerful if they are based on more than one relationship measure.
Generally, the more relationship measures, and the more varied the
relationship measures, the better. With easily observed animals, the rates
of interaction can be calculated for a range of interaction types, including
both affiliative and agonistic behavior, possibly including interactions
in more than one sensory mode, such as calling and grooming. Even for

very cryptic species whose behavior is difficult to observe, it is possible to obtain multiple relationship measures, such as association indices in different behavioral modes (e.g., foraging and socializing; Fig. 4.7) or over different spatial and temporal scales (e.g., using two definitions of association; Section 3.3). With multidimensional relationship data available, the issue becomes how to make best use of them.

We can use the methods of the previous sections of this chapter in tandem. Ordinations, cluster analyses, sociograms, network measures, dominance analyses, and lagged association rates can all be carried out separately for analyses of, say, agonism and grooming, and the results displayed side by side (e.g., Flack et al. 2006). A useful technique introduced by network analysts is to use the same ordination of the nodes representing the individuals for sociograms or network diagrams of different ordination measures. So, for instance, Lusseau (2007) used the association indices within a community of male bottlenose dolphins to arrange the nodes on a network diagram and then used the same arrangement to display the strengths of links of agonistic and affiliative interaction rates (Fig. 5.17). The contrasting patterns indicate that the dolphins were more agonistic to individuals with whom they associated least and more affiliative to individuals with whom they associated most.

Another approach is to produce single displays that incorporate data from different relationship measures. Examples of this are shown in Fig. 4.7, in which the association index of a pair of dolphins when foraging is plotted against the association index of the same pair when socializing, and in Fig. 4.4, in which the dyadic values of four relationship measures for chickadees feeding at a feeder are plotted against one another. With more than two relationship measures, however, as in Fig. 4.4, such displays are rather unwieldy and do not give a clear overall perspective of the social structure. If the relationship measures are to some degree correlated, then principal components analysis (Section 2.6) can be used to reduce dimensionality, perhaps allowing most of the information contained in several relationship measures to be plotted in two dimensions (Whitehead 1997).

As an example, a principal components analysis of the four relationship measures of the chickadees shown in Fig. 4.4 gave two principal components with eigenvalues greater than or equal to one (a frequently used cutoff when deciding how many dimensions to display; Manly 1994, p. 86). The first, which I call "Association," loaded (i.e., correlated) moderately with all the measures except dominance, and the second loaded entirely on dominance, and so I call it "Dominance." Here

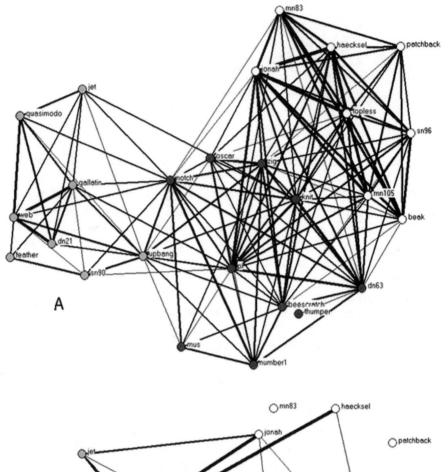

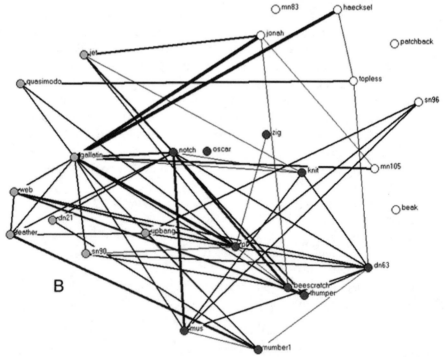

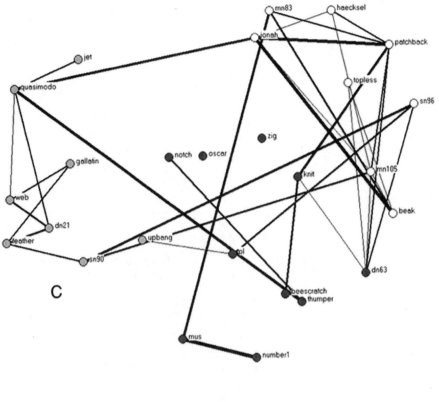

C

FIGURE 5.17 Network depictions of one community of bottlenose dolphins (*Tursiops* spp.) in Fjordland, New Zealand, using the same ordination. (A) Association indices based on grouping. (B) Rates of agonistic "head-butts." (C) Rates of affiliative "mirroring." (From Lusseau 2007) (Illustration copyright Emese Kazár.)

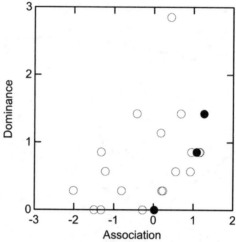

FIGURE 5.18 Principal components scores for four dyadic relationship measures for seven chickadees (*Parus atricapillus*) observed feeding together at a feeder (see fig. 4.4; data from Ficken et al. 1981). The second principal component scores (Dominance) are plotted against the first (Association). Each symbol represents one dyad.

are the loadings (correlation coefficients between original variables and principal components):

	First principal component "Association"	Second principal component "Dominance"
Nearest associates?	0.34	0.00
Dominance	0.00	1.00
Arrived together?	0.41	0.00
Censused in same hour?	0.39	0.00

Because the variables that make up "Association" are all symmetric ($I_{AB} = I_{BA}$), whereas dominance is perfectly asymmetric ($I_{AB} = -I_{BA}$), I only plot values of the second principal component greater than zero against the first principal component in Fig. 5.18 (the part not shown, with "Dominance" less than zero, is a mirror image of that shown). This plot summarizes much of the information available (i.e., in the multiple plots of Fig. 4.4) because the first two principal components together explained 81% of the variance in the data. It clearly indicates that dominance asymmetry increases with association, as well as that the pairs that later mated had high association values but not atypical dominance relationships.

If the relationship measures are categorical rather than continuous, then multiway tables may be a useful alternative display method. Table

Table 5.11 Categorical Summary of Relationships for Ecuadorean Sperm Whales (*Physeter macrocephalus*) for Three Levels of Association Strength[a] and Two Levels of Temporal Stability[b] between Different Classes of Individuals and for Relationships of Female #234

Strength	Stability	Number of pairwise relationships:			
G^a	H^b	♀-♀	♀-♂	♂-♂	♀#234-?
1 (2–4 hr)	0 (unstable)	252	21	0	1
2 (0.1–2 hr)	0 (unstable)	619	45	1	5
3 (together)	0 (unstable)	145	20	0	0
1 (2–4 hr)	1 (stable)	10	0	0	0
2 (0.1–2 hr)	1 (stable)	205	0	0	12
3 (together)	1 (stable)	148	0	0	7

[a]G : 1, seen <4 hr apart but not <2 hr; 2, seen within 2 hr but never together; 3, seen together.
[b]H : 0, not seen together on 2 days at least 10 days apart; 1, seen together on 2 days at least 10 days apart.
Whales are classified as female/immature (♀) or mature male (♂).
From Whitehead (1997).

5.11 shows an example for sperm whale relationships categorized by how close together in time the individuals were identified and over what time period. Using the characteristics of sperm whale society indicated by the summary in Table 5.11, we can see that some relationships among females and immatures are stable, whereas no observed relationship involving a mature male is.

Such displays and tables can be very informative about the nature of a social system (Whitehead 1997). Figure 5.19 displays four different simulated social systems, each containing 30 animals. In these plots, each point represents a dyad, the x-axis represents the mean interaction rate, and the y-axis represents the temporal variability of their interactions (the CV of the interaction rates over days on which at least one member of the pair was observed; Section 4.6). The different social systems produce characteristically distinct plots. We can draw a number of types of inference from such plots (Whitehead 1997):

· *Social complexity*. The complexity of the social structure is indicated by the complexity of the multivariate dyadic relationship measures plot. A simple social structure, with all relationships more or less the same, should produce one cluster of points, with the diameter of the cluster similar to the precision of the position of the points (as in Fig. 5.19a). More complex social structures have more diverse relationships, and so more complex multivariate plots (Fig. 5.19). The possibility of constructing a formal definition of social complexity from this concept is discussed briefly in Section 6.3.

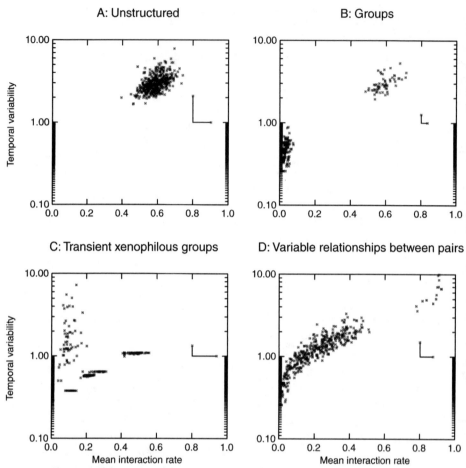

FIGURE 5.19 Representations of the relationships among 30 individuals in four simulated social systems The temporal variability is the coefficient of variation of the interaction rates over days in which at least one member of the dyad was identified. An estimate of the precision of the plots (1.96 times SE for each measure; i.e., half approximate 95% confidence interval) is shown. The four social systems are as follows. **A:** "Unstructured," in which the probability of the dyad interacting during a survey was 0.6. **B:** "Units," in which the probability of interacting during a survey was 0.6 for members of the same unit and 0.02 for members of different groups. **C:** "Transient xenophilous units," in which groups spent 3-day periods in the study area and the probability of interacting during a survey was 0.1 for members of the same unit, 0.8 for members of different units in the study area at the same time, and 0.0 if one individual was in the study area and the other was not. **D:** "Variable relationships between pairs," in which case the interaction rate for a dyad during a survey was chosen from the uniform distribution between 0 and 0.45, except for 15 randomly chosen pairs with rate 0.9 (for further details, see Appendix of Whitehead 1997). (Redrawn from Whitehead 1997, fig. 2.)

- *Classification of relationships.* Sometimes (e.g., Fig. 5.19b, c), the points representing the dyadic relationships fall into clear clusters, which can be used to describe the general pattern of relationships and characterize the social structure. For instance, "There are three types of relationship: pairs who have a permanent strong affiliative relationship, those who have a strong but agonistic relationship, and those who spend little time together and rarely interact." The types could be distinguished using a method of cluster analysis, such as K-means (Section 2.7). Even without clear clusters of points, types of relationship may be identified from regions of the multivariate space. For instance, in Fig. 4.7, the dolphin relationships are labeled "affiliates" who spend most of their time together whatever their behavior (these are mostly mother–calf relationships or males within alliances; Gero et al. 2005), "social associates" who are together about 50% of the time when socializing but much less when foraging, "foraging associates" in which this pattern is reversed (rather few of these dyads), or "acquaintances" who spend little time together in any behavioral mode.
- *Patterns of relationship between classes.* If the plots of dyadic relationship are coded by the classes (such as sex or age-class) of the two individuals, for instance, using colors or marker types (as in Fig. 4.7), then distinctions in relationships between the classes may become apparent.
- *Particular individuals or dyads.* The relationships involving particular individuals or dyads can be highlighted using color, mark type, size, or, in the case of a table, column, and so seen within the context of the social structure (e.g., Fig. 4.4, Table 5.11). With several individuals highlighted, individual variability in social relationships can be assessed.
- *Social units and communities.* Although these multivariate displays do not assume that the population is structured into social units or communities, if it is, then multivariate methods can be used to distinguish them [see Section 5.7 and Whitehead (1997) for more details].

Such plots should be viewed from the perspective of what relationship measures are available as well as the precision of the measures. The society might be highly structured on the basis of a feature (such as a dominance interaction) that has not been measured, and features of the plot that have smaller diameter than the measurement precision can bear no reliable interpretation.

5.7 Delineating Groups, Units, Communities, and Tiers

5.7.1: Divisions in Society. A population may be structured into social entities of animals so that many more interactions occur within social entities than between social entities. This can occur in several quite different ways. I distinguish them using three different terms:

> *Community.* A set of individuals that is largely behaviorally
>   self-contained over all relevant time scales, so that nearly
>   all interactions and associations occur within, rather than
>   between, communities.
> *Group.* A set of animals in mutual association (by some reason-
>   able definition of association) over any time scale (Section
>   3.4).
> *Social unit (or unit).* A set of individuals in (nearly) permanent
>   mutual association, by some reasonable definition of associ-
>   ation.

Thus, if two animals are members of a social unit, they are nearly always grouped, by definition, and if they are members of the same group, then they are very likely to be members of the same community, again following from the definition of community. However, these inclusions do not necessarily work backward: Individuals not of the same unit may form temporary groups, and members of the same community need not be in the same group at any time. There may also be other social entities that cluster associated individuals in some way but do not conform to the definitions of group, unit, or community.

There can be hierarchical levels of social structure, *tiers* in the terminology of Wittemyer et al. (2005), such that the elements of tier i each contains one or more elements of tier $i - 1$. In Wittemyer et al.'s (2005) usage, the upper tiers may be nonsocial aggregations. According to this perspective, in a K-tier society, permanently associating social units, if present, will always be tier 1, the population will be tier K, the community may be tier $K - 1$ (or $K - 2$ if there are subpopulations), and the intermediate tiers may consist of elements, sometimes called "clans," "bond groups," or "pods", within which lower level tiers such as social units show association preference. Such hierarchically structured multi-tier societies are clearly present in elephants (*Loxodonta africana*) (Wittemyer et al. 2005) and killer whales (*Orcinus orca*) (Bigg et al. 1990) and probably many other species.

Populations need not contain units, groups, communities, or tiers, but if they do, these are important elements of the social structure. We would like to assess their sizes and membership. Sometimes these are obvious. If the animals are very visible and the groups, units or communities are well delineated, groups can be counted visually in real time, and if the animals are easily individually identifiable, assessing membership may be equally straightforward. However, with less easily viewable or identifiable animals or situations in which the units, groups, or communities are not so well defined, estimation techniques may be needed. This is especially the case with social units embedded within a fission–fusion society so that any observed group likely contains more than one unit. Appendix 9.5 summarizes some techniques for estimating unit, group, or community size in cases in which individual memberships are not assigned.

If the population being studied consists of multiple social units, groups, or communities, however, it is usually preferable to delineate and assign individuals to them. We can then estimate the sizes of the social entities (just by counting), look at whether the units, groups, or communities contain different proportions of the different classes in the population, whether they behave differently, whether there are different types or rates of dyadic interaction within as opposed to between units, groups, or communities, and many other issues. Sometimes the units, groups, and communities are clear, so that delineating them and assigning individuals is trivial; sometimes it is not. Usually, the reason for lack of clarity is because there is some association between members of different units, groups, or communities. This can occur in a number of ways:

1. Individuals form permanent units, but the units fuse temporarily to form groups.
2. Individuals may occasionally transfer permanently from one unit or community to another.
3. Some individuals float between units and communities or are members of more than one.
4. The spatiotemporal distinction between groups is not very clear.
5. Unit, group, or community distinctions break down in some conditions (e.g., drought or the mating season).

In some cases, two or more of these sources of "social noise" might be present. How do we distinguish the community, group, or social unit signal from the noise?

If the task is dividing a population into groups based on the individuals' locations, which appear to form clusters but not in a clearly unambiguous way (issue 4 in the foregoing list), then there are a number of available techniques as summarized in Section 3.4. These include the "chain rule" and Strauss' (2001) "kth nearest-neighbor hierarchical clustering."

When the target clusters are social units or communities, then it is usually the patterns of identifications and associations (or perhaps interaction rates) by which we wish to delineate clusters. Many of the techniques introduced earlier in this chapter are useful. Sociograms or ordinations using principal coordinates analysis or multidimensional scaling may suggest social units or communities if there are clusters of individuals that are mutually well linked in a sociogram or plotted together in the ordinations (Section 5.2; e.g., Figs. 5.5 and 5.6). However, cluster analyses (Section 5.2), network analyses (Section 5.3), and multivariate methods (Section 5.6) include methodology directed at detecting clusters in the data, and so tend to be more useful. The concept of modularity introduced by the network analysts proves very useful in assessing the utility of divisions. In what follows I give some details on these and other methods. Box 5.6 gives overall recommendations for population division.

5.7.2: Modularity. A particularly useful criterion when assessing the value of a clustering scheme for population division is Newman's (2004) *modularity* as modified for weighted networks. When applied to association indices, modularity is simply the difference between the proportion of the total association within clusters and the expected proportion. The formula for modularity is

$$Q = \frac{\sum_{I,J} \alpha_{IJ} \delta(c_I, c_J)}{\sum_{I,J} \alpha_{IJ}} - \frac{\sum_{I,J} \hat{\alpha}_{IJ} \delta(c_I, c_J)}{\sum_{I,J} \hat{\alpha}_{IJ}} \tag{21}$$

where α_{IJ} is the association index between individuals I and J, $\hat{\alpha}_{IJ}$ is the expected value of α_{IJ} assuming random associations, $\delta(c_I, c_J) = 1$ if I and J are members of the same cluster, and $\delta(c_I, c_J) = 0$ if I and J are members of different clusters.

The modularity Q has expected value 0.0 for randomly assigned clusters and equals 1.0 if there are no associations between members of different clusters. Ideally, we would like to divide the population in a manner that maximizes modularity. Newman (2004) suggests that if $Q \geq 0.3$, then the divisions between clusters are "good."

Newman's (2004) original formulation assumes that the expected association index of a dyad is proportional to the product of the gregariousness of the two individuals, so that

$$\hat{\alpha}_{IJ} = \frac{s_I s_J}{2m} \tag{22}$$

where s_I is the strength of associations, or gregariousness, of individual I ($s_I = \sum \alpha_{I,J}$ summing over J; Section 5.3), $m = \sum s_I/2$ (the total of all dyadic association indices).

D. Lusseau (In preparation) noticed that by changing the definition of the expected association index $\hat{\alpha}_{IJ}$ as it is used in Equation (21), one can achieve different types of community division. Thus, if there are sets of animals that are never identified in the same sampling period, they will tend to be allocated to different clusters using the formula for expected association in Equation (22) [because α_{IJ} is zero between such sets, whereas $\hat{\alpha}_{IJ}$ is not, and thus if $\delta(c_I, c_J) = 0$, with I and J in different clusters, modularity increases]. Alternatively, we can use expected values of association indices calculated in other ways, including from the permutation techniques described in Section 4.9. Here are some possibilities:

1. Not controlling for gregariousness or data structure, so that all dyads have same expected association index $[\hat{\alpha}_{IJ} = 2m/n \cdot (n-1)]$.

2. Controlling for gregariousness of individuals [Equation (22)].

3. Controlling for observed group sizes and number of observations of individuals but not gregariousness (mean of random matrices of association indices when permuting group-by-individual matrix; Section 4.9).

4. Controlling for observed group sizes and number of observations of individuals in each sampling period but not gregariousness (mean of random matrices of association indices when permuting group-by-individual matrix within sampling periods; Section 4.9).

5. Controlling for number of observations and associations of individuals in each sampling period and thus including

gregariousness (mean of random matrices of association indices when permuting matrices of association indices within sampling periods; Section 4.9).

If technique 1, 3, or 4 is used, then gregarious individuals that form large groups will tend to be clustered, whereas less-gregarious animals found in small groups may find themselves in clusters by themselves. In most cases, this will not be desirable, so I have only implemented methods 2 [Newman's (2004) original method, controlling for gregariousness] and 5 (control of data structure and gregariousness) in SOCPROG. I call these modularity-G (for gregariousness) and modularity-P (for permutations). Modularity-G divides populations based on who associated with whom, thus combining association preference with any other tendency to be seen together, whereas modularity-P factors out who could have been seen with whom, focusing directly on preferred or avoided associates. Generally, maximum modularity is higher for modularity-G because identification histories potentially provide a useful way to divide a population. A social analyst could be interested in either or both of these forms of community division.

Although the highest modularity, of whichever type chosen, is a sensible goal of population division, this maximization is not technically straightforward. The next two subsections consider methods for finding divisions in populations with high modularity.

5.7.3: Using Cluster Analysis to Delineate Social Entities. Cluster analyses come in many varieties; a primary division is between hierarchical and nonhierarchical analyses (Section 2.7). Either can be used to divide a population into units or communities. The nonhierarchical version seems to do this more directly. Standard nonhierarchical techniques such as K-means, however, assume a rectangular data matrix (e.g., Table 2.2), allowing the individuals to be thought of as points in Euclidean space. Thus, they are not directly applicable for similarity matrix inputs, as result from association and interaction rate data (e.g., Table 2.5). Hierarchical clustering methods are more commonly used in this context. Dendrograms resulting from hierarchical cluster analyses applied to association data are shown in Figs. 2.4, 5.1 to 5.4, 5.7, and 5.8. Some of these look as though they do contain fairly closed units or communities (e.g., Figs. 5.3 and 5.8), whereas others do not. For instance, Figs. 2.4 and 5.1 represent small, quite well-mixed populations, and Fig. 5.7 is constructed from random data. We need techniques that extract meaningful units or communities in the first cases, but not in the second.

A first consideration is what type of cluster analysis to use. We wish to use methods that are least affected by social processes that mask units, groups, and communities (as listed in points 1–5 at the beginning of this section), as well as by other "noise," such as recording errors. Spuriously high association indices are likely to be produced by (a) unit fusions, (b) transfers, (c) floaters, (d) social breakdowns, (e) recording ambiguities, and (f) recording errors. Single-linkage clustering is vulnerable to these and so is not recommended as a clustering method (Section 5.2). Complete-linkage, which considers the least dissimilar individuals when forming clusters, might be appropriate when tight, socially homogeneous units are being sought, but it tends to produce spurious divisions in more heterogeneous social entities, especially communities. When trying to find units, groups, or communities, I generally recommend the use of average-linkage cluster analysis or perhaps Ward's method. Ward's method maximizes the within-cluster similarity, and although it is designed for Euclidean distances, it can be used on other dissimilarity measures (e.g., Wittemyer et al. 2005).

The second, and most difficult, issue when using cluster analysis to delineate groups, units, or communities is when and if to stop. As we move from the "twigs" along the "branches" towards the "trunk" of a dendrogram, the population is divided into fewer and fewer clusters. Where do we stop, concluding "individuals clustered up to here form units (or communities), individuals not clustered up to this level are in different units (or communities)"? Perhaps we do not stop, concluding that there is no unit or community structure in this population, or we might stop two or more times, delineating several tiers of social organization (Wittemyer et al. 2005). We need a "stopping rule." The stopping rule might be dendrogram wide (e.g., "all links with an average association greater than 0.3 are within units"), or we might examine each join separately, accepting or rejecting it. Stopping rules are also necessary when using nonhierarchical cluster analyses, such as K-means (Section 2.7). To partition a population into clusters, how many clusters should we use?

These are issues encountered in the analysis of many types of data. Much has been written on stopping rules for cluster analysis (e.g., Milligan & Cooper 1987), but, unfortunately, few generally applicable principles have emerged. The "best" technique is very dependent on the situation.

In some cases, it is reasonable to apply a simple rule of thumb based on the social behavior of the animals. Thus, for instance, we might set the association index at 0.5, indicating animals spending half or more of their time together, for defining a level of social structure [as in Bigg et al.'s (1990) "pods"], or twice the mean community-level association index, so that clustered individuals spend at least twice as much time together as a randomly chosen pair (Section 4.9; Table 4.15).

An empirical method described by Wittemyer et al. (2005) examines the structure of the dendrogram looking for "knots"—levels of association such that the rate of cluster formation suddenly changes on either side. This is illustrated in Fig. 5.20, in which the elephant social structure shows two levels of association at which cluster formation decreases. These knots allow the determination of three tiers. Knots can be found by inspecting the knot diagram—the plot of cumulative bifurcations (insert in Fig. 5.20)—and seeing where its shape changes suddenly. Alternatively, Wittemyer et al. (2005) describe more formal statistical methods for placing knots in a dendrogram. When trying out this technique on different data sets, a difficulty I have sometimes encountered is that knots may appear or not appear in different places, depending on the distance metric used on the x-axis of the knot diagram (Fig. 5.20, insert). For instance, one could use the square root of one minus the association index or minus the logarithm of the association index and find different knot values.

Modularity (see prior discussion) provides a guideline for assessing where in a dendrogram to stop clustering. Using a technique developed by Lusseau (2007), one can move up or down a dendrogram, accepting or rejecting bifurcations until Q is maximized. The level of association index giving maximum modularity is chosen. This could be the allocation of all members of the population to the same cluster if no division improves modularity.

As an example of this technique as well as the knot diagram, Fig. 5.21 shows the average-linkage cluster analysis dendrogram, knot diagram, and modularity plot for 54 female and immature sperm whales, using both kinds of modularity. The knot diagram is not very clear, but it might suggest a knot at an association index of about 0.32, whereas modularity-G and modularity-P clearly peak at much lower levels of about 0.1. The modularity-G here of 0.603 is high, showing that with this division, there is much more total association within clusters than would be expected for randomly determined clusters. However, when the identification histories of the individuals are factored out by using permuted associations to estimate the expected association indices, the

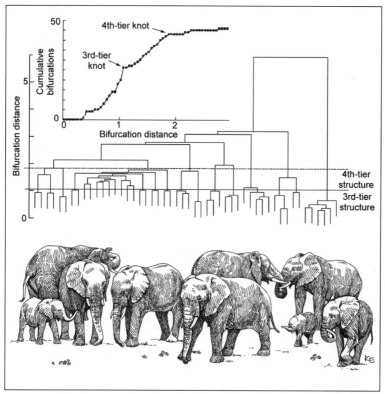

FIGURE 5.20 Dendrogram (Ward's method, cophenetic correlation coefficient [CCC] = 0.82) and knot diagram (insert) for a population of elephants (*Loxodonta africana*) in Kenya. Each node, at the bottom of the dendrogram, is an elephant matriarch, representing the second tier of the social structure. The bifurcation distance on the y-axis of the dendrogram and the x-axis of the knot diagram is inversely related to the simple-ratio association index (with a bifurcation distance of 0 being equivalent to an association index of 1.0, and so always associated). The knot diagram gives the cumulative number of bifurcations at different bifurcation distances. The rate of bifurcation slows considerably at distances of 1.05 and 1.85, the knots. Thus, third-tier structures are formed by bifurcations less than 1.05 and fourth-tier structures by bifurcations between 1.05 and 1.85. (From Wittemyer et al. 2005, fig. 4.) (Illustration copyright Emese Kazár.)

maximum modularity-P is much lower, 0.044 at a cutoff association index of 0.167. Clearly, much of the division within this population is based on different patterns of identification, rather than preferential companionships, within 5-day periods, and only modularity-G meets the criterion of $Q > 0.3$ for a "good" division.

The last two columns of Table 5.12 compare the population divisions using the knot method and modularity-G. The knot method substantially subdivides the clusters produced by maximizing the modularity across the dendrogram.

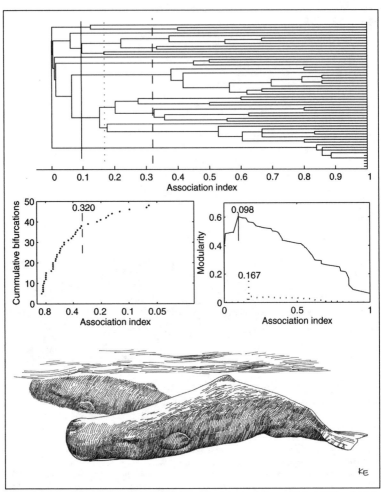

FIGURE 5.21 Above: Dendrogram (average linkage, cophenetic correlation coefficient [CCC] = 0.96) for 54 female and immature sperm whales (*Physeter macrocephalus*) observed off the Galápagos Islands (using sampling periods of 5 days, association being defined as observed within 1 hour, for individuals identified in five or more periods, and a simple-ratio association index). **Below left:** The knot diagram, suggesting a knot at an association index of 0.320 (*dashed line*). **Below right:** The modularity of the dendrogram, suggesting that the best division into clusters is with an association index of 0.098 (*solid line*) when the modularity uses expected values that just correct for gregariousness (modularity-G) and 0.167 (*dotted line*) when the expected values are from permuting associations within sampling periods (modularity-P). (Illustration copyright Emese Kazár.)

Table 5.12 Classification of Female and Immature Sperm Whales Identified off the Galápagos Islands into Social Units Using the Temporal Methods of Christal et al. (1998), the Eigenvectors of the Modularity-G Matrix (Newman 2006), the Maximum Modularity-G of the Nodes on the Dendrogram (Lusseau 2007; Fig. 5.21), and the Knot in the Dendrogram (Wittemyer et al. 2005; Fig. 5.21)

Unit membership	Eigenvector modularity	Dendrogram modularity	Dendrogram: knot
....	....	....	....
....	....	....	
....	....	....	
....	....	....	....
....	....	....	
....	....	....	
....	....	....	....
....	....	....	
....	....	....	....
....	....	....	....
....	....		....
		....	....
....	....	....	....
....	....	....	....
....	....	....	....
....	....	....	....
....	....	....	
....	....	....	
....	....	....	....
....	....	....	
....	....	....	....
....	....	....	
....	....		....
....	....		....
....	....		
????	....	....	....
	....		....
	....	....	....
....	....	....	
....	....	....	....
....	....	....	....
....	....	....	....
....	....		
....	....	....	....
....	....	....	....
....	....	....	....
....	....	....	....
....	....	....	....
....	....	....	....
....	....	....	....
....	....	....	....
....	....	....	....
....	....	....	....

Each row of the table represents an individual, and the heavy horizontal lines divide clusters produced by the different methods.????, an individual whose unit membership could not be determined using the methods of Christal et al. (1998).

The methods described so far find population-wide criteria for stopping when proceeding up the branches of a dendrogram. Another potential approach is to consider each bifurcation separately, accepting or rejecting the linkage using some criterion. I have not seen this done with traditional cluster analysis, but it is implemented in some of the routines of network analysis.

5.7.4: Using Network Analysis to Delineate Units or Communities. Network analysts have long been interested in allocating nodes into communities. In contrast to the hierarchical cluster analysis approaches, which have generally favored an agglomerative method—starting with all individuals separate and progressively clustering them—network theory has emphasized divisive methods in which the population is sequentially divided. A number of algorithms for identifying community structure have been developed, initially for binary networks, but some are functional with weighted networks, such as those based on association indices (Newman 2004).

If the criterion by which a method is judged is modularity, then a new eigenvector-based method of Newman (2006) seems often to be the best. It can work with weighted networks, such as those we encounter in social analysis (e.g., association indices), and successively divides each cluster into two using the positive and negative elements of the dominant eigenvector of the "modularity matrix," adjusting membership of the two daughter clusters to increase modularity at each stage until all further potential divisions reduce modularity. It is fast and easily programmed (e.g., it is included in SOCPROG), and can be applied to modularity-G as well as modularity-P, in which expected association indices are obtained by permutation, although this takes more computation time.

Based on using this method on the sperm whale data, the population was split into six clusters with modularity 0.608 (Table 5.12). This is slightly higher than the maximum modularity-G from the average-linkage dendrogram (Fig. 5.21; 0.603). Similarly, the eigenvector method gives a better division using the permutation method, modularity-P, compared with the dendrogram method: 0.074 versus 0.044. My not-very-extensive experience suggests that the eigenvector method usually does better than, or at least as well as, the dendrogram method of maximizing modularity.

Theoretically, given a metric for assessing the efficiency of any cluster division such as modularity, all possible divisions could be assessed and that giving the highest value chosen. Unfortunately, even with only a few nodes in a network, there are too many potential divisions for this to be practicable. The dendrogram-based method and the eigenvector-based method are short-cuts for finding clusterings that are as nearly optimal as possible. Network analysts will doubtless continue to work at this problem and develop more efficient routines for maximizing modularity or other measures, and their results will be useful to social analysts.

A particularly interesting approach is the division of networks into overlapping communities so that some individuals could be members of more than one community. The sole implementation of this that I am aware of (Palla et al. 2005) works directly only on binary networks, and so is of limited use to those studying nonhuman societies (Section 5.3).

Another fascinating recent development from the network analysts is described by Newman and Leicht (2006), whose technique examines a network without any preconceived notion as to what structure it may have. The likelihood-based analysis clusters nodes on the basis of a wide range of potential similarities in the way they connect to other nodes, including community structure, and its opposite, bipartite structures in which nodes preferentially link to nodes in different clusters. These would correspond to interesting social structures! This technique has another attractive feature: In common with the Bayesian technique discussed later, it gives probabilities that each individual belongs to each cluster. Unfortunately, the original description of this method refers only to binary 1:0 networks.

5.7.5: Using Multivariate Analysis to Delineate Social Entities. A very different approach to finding units or communities comes out of the multivariate techniques described in Section 5.6. The axes in displays, such as those Figs. 4.7 and 5.19, define a relationship space within which each dyad is situated (Whitehead 1997). If a region of this space can be found such that within it relationships are transitive (i.e., if the relationship between A and B and that between B and C are within the region, then the relationship between A and C also lies within the region), then the region can be used to define social units, communities, or other social entities. The condition of transitivity could be relaxed to define semiclosed semiunits or to account for imperfections in the data, for instance, by requiring 80% or more of the pairs of relationships in the region to be transitive (Whitehead 1997).

This approach is broader than most of the others discussed in this section [an exception being the Newman and Leicht (2006) network technique] because it can use two or more relationship measures and be used to define very different sorts of social entity. For instance, the upper-left-hand dispersed cluster of points in Fig. 5.19C is transitive, describes weak but temporally stable relationships, and so could be used to define "loose but permanent units." Asymmetric relationship measures could also be used in this approach, and then regions of space could define dominance hierarchies. A more traditional usage is illustrated in Fig. 4.7: The male–male relationships among the bottlenose dolphins in the upper right quadrants that have high associations both when foraging and socializing are members of "alliances" as described by Connor et al. (1992)].

It is possible to use this approach to define multiple social tiers, which would be represented by two or more different regions of relationship space that satisfy transitivity (Whitehead 1997). In the case of hierarchically arranged tiers (sensu Wittemyer et al. 2005), these regions would enclose one another, but this need not necessarily be the case. There could be some male alliances in one part of the space and female social units in another.

These multivariate methods, although powerful, have rarely been used (but see Whitehead 1997) partly because multivariate relationship data are not often available and partly because there is no standard software for finding regions of transitivity in relationship space.

5.7.6: Temporal Methods for Delineating Groups, Units, or Communities. The methods described in the foregoing operate on summary relationship measures, especially association indices, but such approaches may mask important elements of the available data. For instance, a lagged association rate analysis (Section 5.5) might indicate that individuals form permanent, or nearly permanent, social units but that these associate with one another over shorter time periods. We wish to allocate individuals to the units. Cluster analyses or network analyses using association indices, as described in earlier subsections, are potential techniques, but if two individuals are observed together frequently over a few consecutive days, they will tend to be clustered, even though they may be members of different but interacting units. Thus the clusters produced may or may not reflect true social units. In my experience, in such situations most individuals will be correctly clustered, but a minority will not (e.g., Table 5.12).

An alternative is to use temporal aspects of the data to delineate social units whose general attributes are uncovered by other methods, especially lagged association rates (Section 5.5). The specific method chosen is situation dependent and varies with the results of the lagged association rate analysis, the structure and amount of data, and the certainty of delineation required. There is usually a trade-off between more rigorous methods, for which the resulting unit memberships are very certain but some individuals are not allocated, and looser techniques, in which more of the population is allocated but some individuals may be misallocated. Thus, the method needs to be designed based on the presumed social structure, the data available, and the desired output. Here are two examples from the work of my colleagues.

Ottensmeyer and Whitehead (2003) studied social relationships among long-fin pilot whales (*Globicephala melas*). The standardized lagged association rate fell over periods of 1 hour to 1 week and then stabilized (Fig. 5.14C), whereas a cluster analysis suggested permanent units containing a few animals, but that these units might associate with one another for periods of 1 week or less. To delineate unit membership, Ottensmeyer and Whitehead (2003) first defined "key" individuals identified on at least 4 days, separated from one another by at least 30 days. "Constant companions" of key individuals were considered those identified with the key individual on at least 3 days, separated from one another by at least 30 days, and two or more key individuals and their constant companions were allocated to the same unit if they met this criterion. This method produced seven well-defined units but left many individuals unallocated.

This methodology was based on that of Christal et al.'s (1998) study of sperm whales, which have a similar social structure. Here, units were defined by identifying sets of individuals each of which was identified associated with at least two other members of the unit on at least two occasions separated by at least 30 days (again much larger than the period of association of different units identified by the lagged association rate analysis; Fig. 5.16). In addition, constant companions of key individuals were allocated to units, using the same definitions and methods as with the pilot whales.

These methods seem rather ad hoc, but they are designed specifically with the data and social structure, including its temporal nature, in mind, and so, I believe, produce more reliable allocations than other techniques. For the sperm whale data set, whose dendrogram is illustrated in Fig. 5.21, the unit designation produced by this method agrees generally quite well with that using the modularity-G across the dendrogram or eigenvector modularity-G, although there are some discrepancies (Table 5.12). The knot method makes too fine a division of these data, at least by the standards of Christal et al.'s (1998) unit division.

5.7.7: Bayesian Methods. With very few exceptions (such as the WST dominance rank and cardinal dominance index; Section 5.4), none of the methods considered so far in this chapter starts from a model of social structure and then fits it. They either display the data in what is hoped are meaningful ways (such as ordinations) or fit mathematical models to attributes of the data (as can be done with lagged association rates).

Ideally, we would start with several biological models of social structure, convert these to statistical models, estimate their parameters, and
compare their fits. This is not simple because social models, dyadic by
nature, do not easily translate into tractable statistical models. However,
Durban and Parsons (In press) made a start by constructing Bayesian
models of social units. The method is powerful and informative but
complex to implement.

The social model is that individuals are clustered into social units,
members of which tend to be associated, but may not be, at any time,
and members of different units may sometimes be associated. However,
individuals with similar identification histories (i.e., often seen in same
group) are likely to be members of the same unit. Following the Bayesian
procedure (e.g., Carlin & Louise 2000), prior probability distributions
for the number of units and the probability of any individual being in
any unit are provided by the analyst. These are often, and in the method
of Durban and Parsons (In press) are, "uninformative," so they do not
bias the results in any particular direction. Using Bayes' theorem, one
uses the prior distributions and data to produce posterior distributions
for the number of groups and the probabilities that each individual is
in any particular unit or shares unit membership with any other individual. These outputs are in some respects ideal, conveying not only
the unit structure, but also uncertainty about it. However, obtaining
the posterior distributions in situations like this with many parameters
is challenging. Durban and Parsons (In press) use the Markov chain
Monte Carlo method (Carlin & Louise 2000). With a data set on associations of 83 pilot whales, the method worked and produced useful
results, indicating 16 to 19 social units (95% confidence interval) together with the probabilities of membership of each individual in each
unit (Jankowski 2005). The method was hard to implement, however,
and took many days of computer time to run.

BOX 5.6 *Recommendations for Population Division*

Section 5.7 summarizes a battery of methods that can be used to
divide populations. As a general recommendation for most analysts, I suggest that first ordinations and/or dendrograms (Section
5.2) be produced using association indices or other relationship
measures to see whether there are indications of clustering. If
there does seem to be some sort of population division, then it is
probably worth estimating the modularity using either Newman's

(2006) eigenvector method or the maximum value across the dendrogram, particularly if the dendrogram has a high CCC. If this maximum modularity is less than about 0.3, then the population division is not well supported, and divisions produced using any method should be treated cautiously. If modularity is greater than about 0.3, then the population may be reasonably subdivided. Of the methods that can be used to subdivide populations, using the eigenvector modularity method (Newman 2006) or "stopping" the dendrogram when modularity is maximized seems a good approach, with the former usually performing somewhat better. Neither allocation is necessarily "correct," however, assuming that, from the animals' perspective, there are real units, communities, or other social entities. In addition, only in the case of the eigenvector modularity method with two clusters is there any measure of how well the allocation of a particular individual is supported, and so only in this case can we say something like, "We are pretty sure which unit individual A is in, but much less certain about B."

The analytical procedure that I just outlined can be carried out quite easily with just a few mouse clicks in SOCPROG, and may be as far as most social analysts will go. With more effort, however, considerably more useful results may be attainable:

- Following a lagged association rate analysis (Section 5.5), the temporal patterning of social affiliations may become apparent and permit the delineation of social units using rules designed for the particular data set and the animals' perceived social structure.
- If several relationship measures are available, then the multivariate method of delineating social units or other types of transitive social structure can reveal much greater complexity than the standard univariate methods.
- Finally, fitting true models of social structure using Bayesian methods (or possibly likelihood methods; Section 2.8) can give the most revealing and useful output. I can reasonably hope that future hardware and especially software improvements will make these methods, which are currently hard to implement, available more generally for social analysts.

Table 5.13 Rough Assessment of Utility of Different Techniques for Displaying and Modeling Social Structures for Three General Types of Population

Method	Size of population and number of relationship measures available		
	A: Small, well-studied, several measures	B: Medium, three measures	C: Large, sparse, one measure
List of attributes (section 5.1)	3	3	3
Histograms of relationship measures (section 5.2)	2	4	4
Sociogram (section 5.2)	4	2	1
Principal coordinates analysis (section 5.2)	1	3	2
Multidimensional scaling (section 5.2)	3	3	0
Cluster analysis (section 5.2)	3	3	2
Network measures (section 5.3)	3	3	3
Dominance hierarchy analysis (section 5.4)	4	3	0
Lagged association rates (section 5.5)	2	4	4
Multivariate displays (section 5.6)	4	3	0
Delineating units or communities (section 5.7)	1	3	3

A, small population (~12 animals), with several interaction and association measures collected over a considerable time period; B, medium-sized population (~30 animals) studied over a moderate time period (a few months) with three relationship measures available, one relating to agonistic interactions, one to affiliative interactions, and one to associations; C, a large population (~500 animals) of elusive or cryptic animals studied over a long time period (many years) but with only one association index available. Utility is rated as follows: 4, highly useful; 3, generally somewhat useful or useful in some situations (depends on social structure); 2, occasionally somewhat useful; 1, rarely useful; 0, not feasible.

BOX 5.7 *Describing Social Structure: Recommendations*

In this chapter, I introduced a number of analytical techniques for displaying and modeling social structure. For the social analyst faced with her first data set, the possibilities may seem rather bewildering. Consequently, in this section I suggest some general guidelines for how she might proceed. In Table 5.13, three data sets are envisioned: a small, closely studied population, such as a group of captive primates; a medium-sized population of reasonably accessible animals such as might be the case for fish on a coral reef; and a large, hard-to-study population, epitomized by the pelagic cetaceans. Real populations rarely fit nicely into one of these categories, but interpolation between the columns of

Table 5.13 should give some guidance. The rows of Table 5.13 indicate the general utility of a technique for a population type.

For the small, well-studied population, all the techniques are feasible, but some have limited utility either because they analyze features that may not be present (such as social units or communities), other techniques may be better (nonmetric multidimensional scaling will probably give a better display than principal coordinates with such data), or (as with histograms and lagged association rates) they provide statistical summaries of rather few data points, which can usually be presented separately. Particularly useful for this type of data are likely to be sociograms, analyses of dominance, and, especially, the multivariate displays that can reveal many different kinds of societal attributes.

For the moderate-sized population and data set, all the methods discussed in this chapter may have use (Table 5.13). I singled out histograms of relationship measures and lagged association rates by scoring them "4" rather than "3" on the basis that these general techniques can suggest features, such as social units or hierarchies, that can be investigated using more specialized methods. Faced with this type of data, I suggest that the social analyst first estimate the general attributes of the society (Section 5.1), then try different methods of displaying the relationship measures (sociograms, multidimensional scaling, principal coordinates, or cluster analyses), retaining those that represent the data well on both visual and quantitative (e.g., proportion of variance accounted for, stress, CCC) grounds. These ordinations, together with histograms and lagged association rates, can then be used to assess the presence of general features in the society, such as units and dominance hierarchies, that may be investigated using more specialized techniques.

For large populations that are hard to study, the list of useful methods is narrowed. Many of the displays become unwieldy, and there are no interaction measures with which to examine dominance hierarchies or make multivariate displays. Histograms, lagged association rates, and network analyses become particularly important. Units or communities are often present in such systems, and, if they are, their delineation becomes a priority.

Techniques presented in other chapters of this book may also be useful in displaying and modeling social structures. Examples include the analyses of reciprocity (Section 4.8), preferred

associations (Section 4.9), roles (Section 7.1), and kinship structure (Section 7.3). The allocation of techniques to Chapters 4, 5,
and 7 is to some extent arbitrary.

All these methods are more useful and powerful if class information is included, so that the relationships within and between
the different classes of animal are described and compared.

A final recommendation is that the social analyst stay in-
formed of developments in analyzing nonhuman social structures
by network analysis (Section 5.3) and Bayesian models (Section
5.7). These are potentially powerful techniques under intense development. When these methods have been carefully developed
for this context, validated, and made easy to use, they may become the most preferred methods of social analysis.

6

Comparing Societies

6.1 Comparing Social Structures

Having used the methods of the previous chapters to describe and perhaps to model a social system, it is natural to want to compare social systems. How can this social structure be compared with those of other populations of this or other species? Can we categorize it in some meaningful way? How complex is it? This chapter is about lining social structures up against one another.

A straightforward approach is to measure the same attributes for the different societies and then compare them against one another. The attributes listed in Section 5.1 make a good start. With just a few societies, the measures can be tabulated against one another. As examples, Connor et al. (2000) compare group sizes of different populations of bottlenose dolphins (*Tursiops sp.*) and Maryanski (1987) compares network measures for chimpanzees (*Pan troglodytes*) and gorillas (*Gorilla gorilla*). If a number of social structures are being compared, then a data reduction method such as principal components (Section 2.6) can be used to ordinate the societies.

In an elegant elaboration of this approach, Faust and Skvoretz (2002) fitted the same network structure model to 42 social networks of several species. They then compared each pair of networks, using the ability of the model parameters from network A to explain network B and vice versa as a measure of the similarity between the networks. The matrix of network similarities was input into correspondence analysis (Section 2.6) to ordinate the networks. Unfortunately, to achieve all this, all networks had to be binary (1:0), thus discarding a great deal of the information for the nonhuman societies (Section 5.3).

6.2 Classifying Social Structures

As noted in the introductory chapter (Sections 1.4 and 1.7), scientists have had a propensity for the classification of social structures. In the general terms of all animal social structures, this has not proved useful. In more restricted venues, however, classification has more validity. We might want to categorize the social systems of a taxonomic group [e.g., Kappeler and van Schaik (2002) for primates] or perhaps those that have a particular characteristic (e.g., high-level carnivores). Here are some of the features that have been, or might be, used, with notes on how the methods of previous chapters may play a part:

- *Mating system.* Classification by attributes of the mating system (e.g., monogamy, polygamy, polyandry; Emlen & Oring 1977; Section 7.5) is routine and often useful (e.g., Clutton-Brock 1989). The methods used to determine the mating system are usually based on observations of who mates with whom (Section 3.2), as well as genetic determinations of paternity (Section 4.2).
- *Relationships between sexes.* Societies are also frequently classified by the general relationship patterns between the sexes, so we might have "pair bonds," "single-male groups," and so on (Kappeler & van Schaik 2002). The general pattern of relationships between the sexes can be clearly shown by sociograms, dendrograms, or ordinations (e.g., Figs. 5.3 and 5.4).
- *Number of associates.* Individuals may have zero, one, or many associates, indicating, perhaps, "solitary," "simple," or "diverse" societies. If the y-axis of the histogram of association indices (Section 5.2) is divided by the population size minus one (or the number of animals in the "from" class minus one in the case of division by class), then this indicates the number of associates an average individual has at different strengths of association. This is done in Fig. 6.1 for the

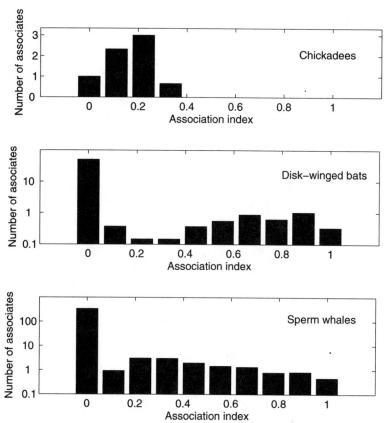

FIGURE 6.1 Histograms of the mean numbers of associates at different levels of association index for chickadees (*Parus atricapillus*) at a feeder, roosting disk-winged bats (*Thyroptera tricolor*), and sperm whales (*Physeter nacrocephalus*) off the Galápagos Islands (for more details, see captions of figs. 5.1, 5.3, and 5.4, respectively).

chickadee, disk-winged bat, and sperm whale data whose original histograms are shown in Figs. 5.1, 5.3, and 5.4, respectively. The chickadees have quite low association rates with all other members of their small community and the bats tend to have about three close associates (association index >0.5), whereas the sperm whale society seems more diverse, with the average member possessing about 10 to 20 moderate or strong associates.

· *Stability of associations.* The term fission–fusion is often used to delineate a class of society in which groups of various characters form and reform frequently (Conradt & Roper 2005). The alternative is a social system in which there is little temporal change in associations, at least over short time periods. The distinction may be important in several respects. For instance, it is argued that in fission–fusion societies there is greater selection for cognitive abilities (Barrett et al. 2003). Fission–fusion societies themselves might also be usefully subdivided depending on whether the groups are formed within closed communities or an open population, or whether the fission and fusion occur within hierarchically nested tiers of social entities (C. Garroway, personal communication). The presence of fission–fusion is clearly indicated by a fall over short time periods in the lagged association rate or standardized lagged association rate (Section 5.5).

· *Compartmentalization or modularity of social structure.* One of the ways of categorizing fission–fusion societies is in the manner of their division. Are there closed communities or a more open system, with individuals possessing a huge range of potential affiliates? In this context, methods of finding optimum modularity are especially useful, such as Newman's (2006) eigenvector method, which can indicate that the best division is no division. Another dimension of modularity is whether there are hierarchically nested tiers of social entity (Section 5.7; Wittemyer et al. 2005).

Some of the other elements of Wilson's (1975, pp. 16–18) list of 10 "qualities of sociality" (Section 1.8) might provide useful ways to categorize societies, as may the list of measures of social structure given in Table 1.2. However, I think that, except when there are clear distinctions, social structures are better displayed and measured (Chapter 5) rather than categorized.

6.3 How Complex Is My Society?

In this section, I consider the concept of social complexity, and begin by discussing the possibility of a lack of sociality and then the simple situation in which an individual's associations are not differentiated. Many types of societies have differentiated relationships, but can they be ranked in complexity? The final subsection addresses this challenging topic.

6.3.1: Minimum Conditions for Sociality? A baseline minimum condition for sociality is that the members of a population interact behaviorally.

If there are no interactions, so that no behavior of any animal has a direct effect on the behavior of any other, then there is no sociality. Almost all sexually reproducing animals interact, and so, when trying to assess the presence of sociality, we might want to exclude mating interactions. We might also want to exclude negative or agonistic interactions, so that if the only way that two individuals respond to each others' presence is to avoid or fight with one another, then the population is not considered social.

Another approach toward asociality that is more generally feasible but also has challenges is using associations. If association is defined using circumstances in which an interaction might occur (Section 3.3), then a pair could be in such circumstances but not interact. For instance, aggregations of individuals in space and time may result either from sociality or from some response to a common external factor, such as resource distribution (Section 3.4).

As an example, consider the humpback whales (*Megaptera novaeangliae*) that spend the summer months off the coast of Newfoundland. Although the waters are wide, the whales are clustered: Groups of a few animals swim a body length apart from one another at a distance of tens, hundreds, or thousands of body lengths from the nearest neighboring group. This seems clearly social—but maybe not. The group sizes of the humpbacks are closely related to the horizontal extent of the prey schools on which they feed ($r = 0.603$, $P < 0.01$; Whitehead 1983), and permutation tests rejected the null hypothesis of preferred or avoided companionship (Whitehead et al. 1982). Thus, perhaps group formation in humpbacks is entirely a result of animals aggregating at prey schools. However, when attention turns to resting humpbacks, for whom there is no important spatially structured resource, I found a preference for pairs. Thus, at least when at rest, the humpback whale is social (Whitehead 1983).

It will be impossible in most circumstances to conclude that a population is not social, using any reasonable definition of asociality. However, we can perhaps say that there is no evidence of sociality, for instance, if all observed interactions have to do with mating or are negative (animals avoiding one another) or agonistic, or all observed associations are simple aggregations associated with resource clusters.

6.3.2: The Null Model: Equivalence. Null models are useful in social analysis, as in other areas of science, when investigating potentially complex phenomena. They allow us to formalize, at least partially and in a negative way, what we mean by complexity. If the null model is rejected, we can conclude that "the society is not one whose relationships can be explained by simple SOMETHING." But what is the SOMETHING?

Equivalence is one possibility. The concept of equivalence was originally developed in the context of linguistic processing, but Schusterman et al. (2000) suggest that it may have value in social analysis. From a particular individual's perspective, some of its social partners may be considered equivalent. As Connor et al. (2001) point out, Bejder et al.'s (1998) permutation test for preferred/avoided associates and its variants (Section 4.9) can test the equivalence null model. In their application, Connor et al. (2001) showed that within a "superalliance" of 14 male bottlenose dolphins (*Tursiops* spp.), the null hypothesis of no preferred/avoided companions during consortships with females (i.e., equivalence of all other members of the superalliance in this context) could be rejected using the Bejder et al. (1998) test, thus implying a complex internal structure to the superalliance.

At its most basic, this type of analysis could be carried out with the entire population, thus testing whether there are any preferred/avoided companions or differentiated relationships. If this test does not reject the null, equivalence, hypothesis, then further analysis of social structure (as described in Chapter 5) makes little sense. Obviously, the probability of rejecting the null hypothesis depends on the amount of data available, and in subtly structured societies equivalence may not be rejected if data are rather few (Section 3.11). Thus, obtaining a P-value greater than the critical level in the Bejder et al. (1998) test does not necessarily imply that there is no social structure, but rather that there are insufficient data to characterize it.

6.3.3: Quantifying Social Complexity. There is a range of reasons for trying to quantify social complexity. We may be interested in the phenomenon itself: What are its characteristics, where and how has it evolved, what are its limits? The potential evolutionary and ecological links between social complexity and other characteristics of species and populations have provoked a great deal of speculation and some quantitative research. Are socially complex species more ecologically successful (Rendell & Whitehead 2001)? Are they more cultural, and if so, did social complexity beget complex culture (Roper 1986, but see Reader & Lefebvre 2001) or perhaps vice versa (Richerson & Boyd 1998)? In particular, in the most high-profile controversy of all, did social complexity drive the evolution of large brains and increasing cognitive abilities in primates and other groups of animals, the "Machiavellian intelligence" or "social brain" hypothesis (Humphrey 1976; Byrne & Whiten 1988; Dunbar 1998; Whiten 2000)?

To examine such questions, one needs to quantify, or at least rank, social complexity. But this is difficult. Few researchers have gone beyond community size or group size (e.g., Marino 1996; Dunbar 1998). These have some relevance because they indicate the number of social partners of an individual, with community size referring to the number of different potential interactants of an individual at any time and group size referring to the mean number of actual interactants. However, surely, from the perspective of an issue like the "social brain" hypothesis, we should consider how dyads interact, not just whether they might. If all potential or actual interactants are considered "equivalent" (Schusterman et al. 2000; see prior discussion) and not individually discriminated, then in its functional significance and cognitive representation, the group of animals in which an individual is included might be comparable to a tree that it uses for shade. In contrast, if each dyadic relationship is characterized by a particular pattern of interactions, then society is much more complex, even though community or group sizes may be similar.

In consequence, there has been a search for better measures of social complexity. This has proved to be difficult. Potential conceptual dimensions of social complexity are presented by Whiten (2000) from the perspective of examining the Machiavellian intelligence hypothesis: levels of social structure (including roles and tiers), dyadic complexity (which includes gregariousness and reciprocity), polyadic complexity (complexities of interactions involving more than two participants), variability of response, stability of relationships, complexity of prediction of behavior of others, and demographic complexity.

But how can we operationalize these? To simplify matters, let us first consider one relationship measure, the interaction rates of some activity or an association index, summarized by a square matrix (e.g., Table 2.5 or 4.4). My ideal would be to find a measure that (1) indicates the degree to which individuals in a population prefer or avoid the companionship of, or preferentially interact with, other particular individuals (2) is approximately unbiased by the population size or features of the sampling regime, and (3) is virtually unchanged by the addition into the study population of separate communities. I do not know of such a measure, but the methods that have been developed for social analysis and described in the previous chapters indicate several partially satisfactory candidates and possible directions for discovery. They are listed here with advantages and disadvantages:

- *Group size or gregariousness.* Typical group size or gregariousness (Section 4.3) indicates the mean number of individuals that an individual may interact with at any time, but they do not say whether they do interact, how they interact, or whether the interactions among different partners are in any way different.
- *Community size.* Community size indicates the maximum number of individuals that an individual might interact with more than very occasionally, but it does not say anything about diversity in the types or rates of interactions.
- *Social differentiation.* Social differentiation (Section 5.1), the estimated coefficient of variation of the true association indices in a population, indicates how rates of association vary among dyads. It satisfies the first two of my criteria because it indicates the heterogeneity of associations and is designed to be approximately unbiased by sampling intensity but not the third criterion: A population consisting of two barely interacting but internally homogeneous communities will have a much higher social differentiation than a population consisting of just one such community.
- *Within-community social differentiation.* To get around the failure of social differentiation to satisfy criterion 3, populations could be delineated into communities using the methods summarized in Section 5.7, social differentiation estimated for each community, and then the mean of these used for the population. This seems to satisfy all criteria, at least approximately, but may be heavily dependent on the criteria used for community delineation. Different assignment techniques can produce quite different communities (Section 5.7; Table 5.12), which in turn might give quite different estimates of within-community social differentiation for the same population.
- *Observed/expected measures.* The Bejder et al. (1998) permutation method and its variants (Section 4.9) provide expected values of social measures assuming no preferred/avoided associations, but given the general structure of the data. We (L. Bejder, D. Fletcher, and H. Whitehead, unpublished) have made some preliminary examinations of whether the ratio of the observed to expected values of statistics of social diversity (such as the CV of the association indices) would provide useful measures of social complexity.

Unfortunately, we found no such measure that had reasonable performance under my criterion 2, a lack of bias from population size or the nature of the sampling regime.

- *Information/entropy measures.* The Shannon–Weaver (or Shannon–Wiener) measure of entropy or information (Shannon & Weaver 1949) is used in disparate areas of science to quantify diversity. If we applied it to the association indices of an individual, it would increase with the number of associates, as desired, but decrease with the diversity of its association indices with others, the opposite trend to that implied by my criterion 1. Thus, it is not directly applicable to this situation, although some modification might prove useful.

- *Disparity.* Of the network measures introduced in Section 5.3, disparity (Barthélemy et al. 2005), measured for an individual as the sum of the squares of its association indices divided by the square of its gregariousness, seems on the surface to be the most promising approach to a measure of social complexity. However, as expressed in Equation (15), disparity is highest (1.0) for an individual with just one associate and generally decreases with the numbers of associates, not a desirable attribute for a measure of social complexity.

- *Multivariate diversity.* Multivariate displays of social structure, such as those in Fig. 5.19, indicate the diversity of social relationships well (Section 5.6). It might be possible to quantify this diversity in some way (e.g., using multiscale entropy based upon wavelets; Starck & Murtagh 1999), but could this be done in such a way so that one could make robust comparisons between populations that were sampled in different ways with different intensities and might possess very different structures? This would be very challenging.

- *Diversity of roles.* Blumstein and Armitage (1998) assigned individual rodents to discrete roles (see Section 7.1) using demographic data (e.g., "adult female", "subadult male"), and then measured at the diversity of these roles within communities of different species as measures of social complexity. In this approach, demographic differences are used as proxies for differences in social behavior, and the diversity of social interactions within and between demographic roles is not addressed.

None of the possibilities that I have examined is fully satisfactory as a measure of social complexity. The best, so far, seems to be the within-community social differentiation, but this suffers from the lack of a robust and accepted way of delineating communities. One can hope that there will be more work in this area, and some of the ideas outlined here, such as network statistics, information, or multivariate diversity, might provide starting avenues for other approaches.

As a final caution when considering the topic of social complexity, I note that it is becoming increasingly clear that complex social behavior can arise from just a few behavioral rules of individuals (Sumpter 2006). Thus, a beautiful, complex society may just be an emergent property of automatons. By trying to measure diversity at the level of dyadic interactions or associations among identified individuals, however, as in some of the approaches outlined above, we might be able to avoid this trap.

7   What Determines Social Structure, and What Does Social Structure Determine?

In this chapter, I shift emphasis more to the perspective of the behavioral ecologist who, working from the paradigm of Darwinian evolution, wants to know why behavior has the form it does (Section 1.5). The behavioral ecologist has three main sets of tools: experiments, comparisons between individuals of the same population or species, and comparisons between species (Krebs & Davies 1991, pp. 24–25). A supplementary technique, which is perhaps especially useful for generating and validating reasonable hypotheses, is mathematical modeling (Grafen 1991).

Behavioral ecologists ask many questions about animal societies. I discuss some of them here, focusing on factors that may affect social structure and may be affected by it. Most of them roughly correspond to the upper circles representing "independent or intervening factors" in Fig. 1.4 [Hinde's (1976) framework]. The factors are individuals and their roles (Hinde's "status"), dyads ("effects of interactions on interactions"), environment, the mating system ("age/sex classes"), and culture ("cultural institutions").

The depth and comprehensiveness of coverage in this chapter is generally less than those in its predecessors. There are many ideas about how social systems relate to other parts of biology, evolutionarily, ecologically, or in other ways. Much has been written in some of these areas, for instance, about mating and social systems, and the quantitative methods used at this stage of investigation are usually more straightforward than when examining relationships or describing social systems (for a summary, see Box 7.1).

My focus is on how the methods of analyzing relationships and societies described in Chapters 4 through 6 can be used in the examinations of these issues. In this context, the behavioral ecologist's toolbox is modified a little; the principal techniques used are comparisons among populations and species, comparisons among dyads, and experiments. Variation in social measures among populations of the same species or the same population at different times can be very instructive, and the unit of analysis in social studies is often the dyad rather than the individual.

7.1 The Individual in Society: Roles

Role theory is one of the important paradigms used to examine human societies, as well as those of some nonhumans (Roney & Maestripieri 2003). Roles are fundamental to the societies of the social insects, but they also have a significant influence on the social structures and other aspects of the biology of some vertebrates. For instance, McComb et al. (2001) found that the experience of an elephant matriarch determined the appropriateness of the response of the social unit when faced with a potential threat (Fig. 7.3), and Flack et al. (2006) removed individuals with an apparent policing role from a macaque (*Macaca nemestrina*) society and observed the social structure change. There are two principal elements to such studies: the identification of roles, and the sometimes experimental examination of how roles affect societies. These are the subjects of the next two subsections.

7.1.1: Identifying Roles. To assign roles to individuals, we need one or preferably more measures of behavior. Sometimes, these measures are of individual behavior. For instance, Fresneau and Dupuy (1988) identified all members in a captive colony of 36 ponerine ants and noted the activity of each member on 85 occasions using 12 behavioral categories, such as "nest exploration," "care of larvae," and "foraging." They thus had a 36 × 12 table of counts of individuals performing each activity. They analyzed the data using correspondence analysis (Section 2.6) and cluster analysis (Section 2.7). Figures 7.1 and 7.2 show the resulting displays. Both show the division of the population into sets of animals with distinct roles. The dendrogram from the cluster analysis is perhaps clearest, but the correspondence analysis ordination (which represents

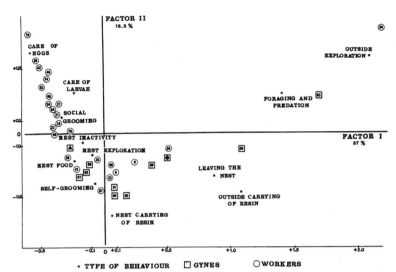

FIGURE 7.1 Ordination resulting from a correspondence analysis of the activities of members of an ant (*Neoponera apicalis*) colony. Numbers represent different individuals, and the closed triangle and closed circle represent the queens. Loadings of activity measures are indicated by asterisks. (From Fresneau & Dupuy 1988, fig. 1.)

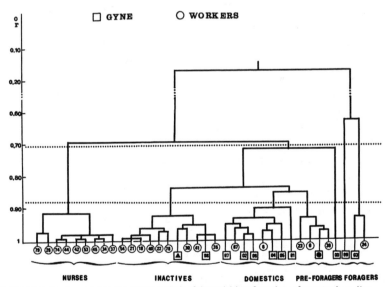

FIGURE 7.2 Dendrogram from the cluster analysis of the activities of members of an ant colony. Numbers represent different individuals, and the closed triangle and closed circle represent the queens. (Fresneau & Dupuy 1988, fig. 3.)

76% of the variance in the data set) also gives information on how the roles line up with the measured activities.

In vertebrate societies, it is more usual to assign roles based on dyadic behavior than individual behavior. Thus, for instance, Flack et al. (2006) distinguished policers in a macaque society as those receiving a disproportionate number of "silent-bared teeth" subordination displays. More generally, this method involves looking at the distribution of some, usually asymmetric, interaction measure and finding sets of individuals with distinctive values, such as those who groom disproportionately often or win the large proportion of agonistic encounters (the "alpha" individual). This can be extended to several interaction measures, such as those individuals who both groom particularly frequently and are generally submissive.

Roles can also be assigned based on associations. In taking this route, the tools of network analysis (Section 5.3) are particularly useful. For example, Lusseau and Newman (2004) described bottlenose dolphins (*Tursiops* spp.) with high betweenness as having the "role of brokers between communities."

Moving beyond dyadic behavior, individuals may have a role in group behavior. This is well illustrated by the significance of matriarchs in elephant society. The matriarch is both agonistically dominant among the females in a social unit and an altruistic leader of the unit (Payne 2003). She makes decisions for the unit and leads their movements (McComb et al. 2001; Payne 2003).

A caution: For any measure calculated for members of a community or population, whether of individual, dyadic, or group behavior, there will be individuals with relatively high and low values. Before assigning roles based on such extremes, we need to know whether they are particularly high or low given the expectations for a randomly interacting or associating, or just behaving, society. Thus, unless the categorization is particularly obvious, the techniques discussed in Sections 4.3, 4.8, and 4.9 for examining distributions of individual and dyadic behavior should be considered before assigning roles.

7.1.2: How Do Roles Affect Societies? If individuals have roles in societies, there are a variety of techniques that can be used to assess the importance of these individuals in the social structure. The basic idea is to compare situations with and without such individuals. The most informative method will usually be experimentally to remove the individuals in question. Examples are Singh and D'Souza's (1992) removal of an alpha male from a captive community of Japanese macaques (*Macaca fuscata*) and Flack et al.'s (2006) "experimental knockout" of policing macaques from a colony. However, experimental removal will not be practical, ethical, or legal in many field situations. A less definitive alternative is the "topological knockout" in which measures of a community's social behavior are compared between the full data set and with the data involving certain key individuals removed (e.g., Table 5.4). Another approach is to compare attributes of social elements (groups, units, communities, or populations) that naturally vary in the presence or qualities of the individuals with a particular role.

Once the contrasting social settings have been set up, we need response measures. These can vary all the way from reproductive success (e.g., McComb et al. 2001), through network measures of dyadic behavior (e.g., Flack et al. 2006), to interaction rates (e.g., Singh & D'Souza 1992). Appropriate statistical techniques depend on the experimental design and may range from general linear models when comparing societies with different attributes to paired t-tests in the case of experimental knockouts [see, e.g., Ruxton and Colegrave (2006) or Sokal and Rohlf (1994) for summaries of statistical techniques for analyzing experiments and Box 7.1].

An example of this approach is McComb et al.'s (2001) comparison of group behavior and reproductive success among elephant units whose matriarchs possessed different levels of experience. A logistic model (in which continuous and categorical variables may affect the probability of an event occurring) was used to examine the influence of various factors on the probability that a family of elephants "bunched" in response to the playback of the distinctive sounds of an adult female from another family. Several factors had little effect on bunching: the mean age of females in the group excluding the matriarch, the number of females in the group, the number of calves in the group, the age of the youngest calf in the group, and the presence or absence of adult males. These were removed from the logistic model. Remaining as factors that affected the probability of bunching were the family being observed, the age of the matriarch of the family being observed, the simple ratio association index (Section 4.5) between the family being observed and that of the family of the individual whose sounds were being played back, and the interaction between the matriarch's age and the association index. These results showed that families with older matriarchs bunched less and displayed a more appropriate response by not bunching in response to calls from familiar individuals than those with younger matriarchs (Fig. 7.3).

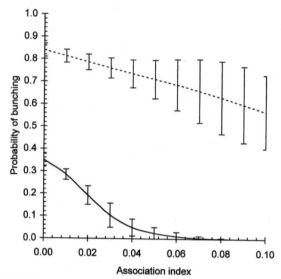

FIGURE 7.3 Probability of bunching by families of African elephants (*Loxodonta africana*) in response to playbacks of calls from individuals of varying familiarity, as indicated by the simple-ratio association index on the x-axis, for families with (*solid line*) 55-year-old matriarchs and (*dashed line*) 35-year-old matriarchs estimated using logistic regression. (From McComb et al. 2001, fig. 1A.)

7.2 *The Dyad in Society: Conflict*

The Darwinian paradigm views organisms as competitors for natural resources. Those that obtain the most, whether it be food, mates, or territory, have the most surviving offspring, and so genes that code for success in these competitions propagate. The consequence is that we expect the behavior of organisms to be adapted for efficient competition; those whose behavior was not well adapted will, statistically and evolutionarily, have perished. From this perspective, conflict is to be expected. There will conflict over food, mates, and territory, and conflict should structure societies.

Thus, agonistic interactions between dyads are to be expected. Efficient agonism should have evolved and should generally be adaptive. Much of behavioral ecology is about how individuals interact agonistically, who wins, how much energy they should devote to a contest, what risks they should take, and so on. From a societal perspective, we are interested in how this agonism socially structures a community. As discussed in Section 7.3, competition may induce cooperation, but if it does not, then the society is relatively simpler. The important issues are what is being competed for and how the competition takes place.

Some of the methods and measures introduced in the previous chapters can help when investigating the role of agonistic interactions in societies. Individuals can be characterized by rates of agonistic interactions, dominance rank, and dominance indices (Sections 4.3 and 5.4). At the dyadic level, asymmetry can be measured using several methods (Section 4.8), including de Vries' et al.'s (2006) dyadic dominance index [Equation (10)]. The most useful societal measures include the overall rates of agonistic encounters, rates within and between classes, Landau's (1951) and de Vries' (1995) indices of dominance linearity, and Vries et al.'s (2006) steepness (Section 5.4). These give rates of overt conflict and how linear and how predictable outcomes of conflicts are. The behavioral ecologist can then use these measures as input for her intraspecific, interspecific or experimental analyses. I give an example of each of the three principal methods being used to examine the role of conflict in structuring society.

At the individual level, rates of agonism, dominance rank, or dominance indices can be compared with age, mass, or other characteristics to look for determinants of agonistic behavior. By making dominance the dependent variable, one can examine how reproductive success varies with dominance, directly addressing the ultimate currency of the behavioral ecologist. Copulation frequency is not quite reproductive success, but for polygynous male mammals, they are closely correlated. Copulation frequency is plotted against rank in Fig. 7.4 for male northern elephant seals. In 1968, when there were 193 females and 103 males at the colony, the most dominant males gained a huge proportion of the copulations, whereas 5 years later, when the colony had grown to 470 females and 180 males, the reproductive success was less skewed (le Boeuf 1974). Similar issues can be investigated at a dyadic level. How does the agonistic interaction rate of a pair or their dyadic dominance index depend on their difference in age, mass, or kinship?

Sometimes, experiments are feasible, usually with captive populations. For instance, Flack et al. (2005) removed the three highest-ranking males from a captive population of 84 pigtailed macaques on randomly chosen days and compared interaction measures and network structure within the remaining community between days with and without the removal. The rate of aggression increased 30% (t-test, $P = 0.02$) without the three "policers."

As an example of the use of the method of comparing species, Isbell and Pruetz (1998) found a statistically significant index of dominance linearity of 0.79 in vervet monkeys (*Cercopithecus aethiops*) but an index of 0.21, not significantly from random, in patas monkeys (*Erythrocebus patas*)

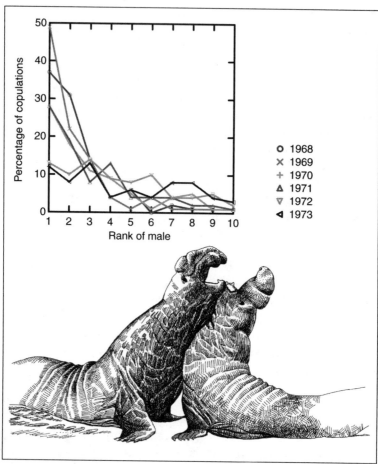

FIGURE 7.4 Number of copulations observed versus dominance rank for males in a colony of northern elephant seals (*Mirounga angustirostris*) breeding on Año Nuevo Island off California over 6 years (Data from le Boeuf 1974, table 5.) (Illustration copyright Emese Kazár.)

at the same site. They attribute the difference to usurpability of food. When food is usurpable—able to be taken over—agonism pays and dominance hierarchies build up. When food cannot usually be removed from its discoverer or contested for in other economical ways, then societies become more equitable (e.g., Kappeler & van Schaik 2002).

7.3 The Dyad in Society: Cooperation

Whereas the Darwinian expects conflict, cooperation between animals is more intriguing, and behavioral ecologists have looked deeply at the mechanisms and evolution of cooperation. The principal processes that have been invoked to explain cooperative behavior are mutualism, kin selection, and various forms of reciprocity. In this section, I discuss each of these mechanisms, using quantitative examples to illustrate potential methodology. Culture can also promote cooperation (Section 7.6). The final subsection considers the behavioral underpinnings of bonds between animals. Statistical methodology is summarized in Box 7.1.

7.3.1: The Puzzle of Cooperation: Mutualism. Mutualism is, in some respects, the simplest of the cooperative mechanisms. Animals behave in ways that help others because the behavior also helps themselves. Grouping is the prime example of mutualism. Animals form groups because it benefits them, and by joining a group, the other members of the group are benefited. Thus, mutualism may structure societies through group formation. Behavioral ecologists examine the benefits of grouping in a variety of ways, occasionally experimentally by changing group sizes (e.g., Williams et al. 2003), but more normally by comparing the costs, for instance, in predation rates, and benefits, for instance, in food obtained per unit time, faced by animals in different-sized groups. Packer et al.'s (1990) study of female lions provides a detailed example of this approach. In essence, when food was abundant, the lions had similar rates of energy acquisition at group sizes of one to four, whereas with food scarcity, the lions did much better either alone or in larger groups of five to seven animals (Fig. 7.5; with abundant food, no significant differences; with scarce food, null hypothesis of equal food acquisition rates rejected at $P < 0.02$; Mann-Whitney U-tests).

Whether the groups formed through mutualistic mechanisms are ephemeral or permanent units may well depend on how the benefits are accrued. If they depend simply on numbers, as in fish schooling to combat predation (Pitcher 1986), then the groups are likely to change composition, whereas if they depend on more complex behavioral coordination, then permanent membership may be beneficial. When individuals do form permanent units, however, it is very likely that other mechanisms, such as reciprocity or, especially, kinship are also in play (e.g., Packer et al. 1991).

7.3.2: The Puzzle of Cooperation: The Role of Kinship. *Kinship* is considered to have a major influence on social relationships (Hamilton 1964) and thus on social structure. The theoretical significance of kinship comes from the genetic basis of evolution through natural selection. Because kin carry some of the same genes, genes that promote behavior in an individual that aids its kin will likely flourish, as will the kin-helping

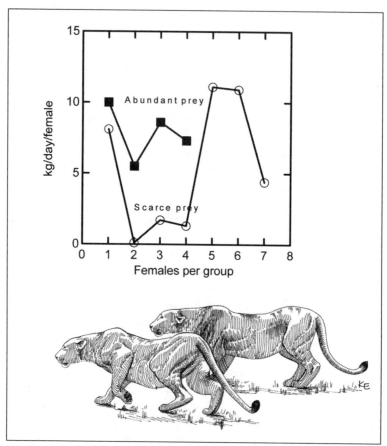

FIGURE 7.5 Daily rates of food acquisition in groups of female lions (*Panthera leo*) of different sizes during times of food abundance and food scarcity. (Redrawn from Packer et al. 1990, fig. 1.) (Illustration copyright Emese Kazár.)

behavior. Hamilton (1964) formalized this process by showing that behavior that increases inclusive fitness—the fitness of an individual plus the fitness of each kin multiplied by its genetic relatedness (the probability that two individuals share a gene through descent)—will be selected for. Thus, we might expect affiliative interaction rates to increase among kin, agonistic interaction rates to decrease, and association indices to be higher.

To test these hypotheses and examine the sizes of the effects if present, we need information on kinship between members of the population. Occasionally, this comes from knowledge of the genealogy of the studied population, but usually this unknown, especially on the paternal side.

Much more frequently, genetic relatedness is estimated from molecular genetic data, particularly microsatellites, and often using the methods of Queller and Goodnight (1989), whose KINSHIP program is frequently used to estimate relatedness (Section 2.9). Table 4.2 shows an example of a table of estimates of dyadic relatedness.

With a relatedness table like this, the hypotheses of kin selection can be translated into statistical tests. The most direct and usual is the Mantel test (Section 2.4). It tests whether there is a linear relationship between the elements of the two identically indexed (usually by individuals in the population) similarity or dissimilarity matrices. Alternatives to the null hypothesis of no relationship when testing for the role of kinship will usually be that there is a positive relationship between kinship and association indices or affiliative interaction rates and a negative relationship between kinship and agonistic interaction rates. Variants on the Mantel test (Section 2.4) may be appropriate. There are nonparametric versions in which elements of the matrices are replaced by their ranks, and partial Mantel tests in which the relationship between two variables is controlled for a third (Smouse et al. 1986; Hemelrijk 1990a). These could be used when studying kinship, for instance, to see whether association increase with relatedness, controlling for habitat overlap.

As an example of these methods, Table 7.1 shows matrices of association indices (home range overlap) and relatedness for greater horseshoe

Table 7.1 Pairwise Relatedness Values (above Diagonal) and Home-Range Overlaps (95% Kernels; below Diagonal) of Greater Horseshoe Bats (*Rhinolophus ferrumequinum*)

	#9684	#9837	#10930	#9668	#10034	#10361	#10758
#9684		−0.05 (0.25)	0.28 (0.5)	−0.06 (0.05)	0.32 (0.05)	0.38 (0.05)	0.59 (0.05)
#9837	0.58		0.13 (0.125)	0.14 (0.05)	−0.07 (0.05)	0.01 (0.05)	−0.38 (0.05)
#10930	0.61	0.51		−0.24 (0.05)	0.13 (0.05)	0.12 (0.05)	0.41 (0.05)
#9668	0.31	0.37	0.29		0.68 (0.5)	0.30 (0.25)	−0.25 (0.125)
#10034	0.33	0.38	0.28	0.65		0.32 (0.5)	0.06 (0.25)
#10361	0.3	0.42	0.42	0.63	0.57		0.40 (0.5)
#10758	0.46	0.65	0.54	0.39	0.42	0.45	

Relatedness was calculated using seven microsatellites and methods of Queller and Goodnight (1989), together with, in parentheses, Hamilton's (1964) theoretical coefficients of relatedness based on known genealogy. Matrix correlations and P-values based on Mantel tests for the relationship between relatedness and home-range overlap were $r = 0.27$ and $P > 0.05$ for microsatellite-based relatedness and $r = 0.64$ and $P < 0.05$ for genealogy-based relatedness. From Rossiter et al. (2002).

bats derived using both known genealogy and molecular genetics (Rossiter et al. 2002). In this case, the significant positive relationship predicted by kinship theory was found using the genealogical measures of relatedness, but although the correlation calculated from the molecular genetic relatedness was positive, it was not significant. This may be because seven microsatellites were insufficient to give accurate measures of relatedness in this species (Rossiter et al. 2002).

Archie et al. (2006) looked at how well genetic relatedness predicts association across several tiers of elephant (*Loxodonta africana*) society. They found positive matrix correlations, and significant results of a Mantel test, between genetic relatedness (estimated using 11 microsatellites) and association (calculated using a 100-meter chain rule) among females in 8 of 10 "core groups" (what I call social units). They also used a Mantel test to show that core groups that shared the same mitochondrial haplotype, and so were more likely to be matrilineally related, associated more than those that did not. In this case, the relatedness matrix was 1:0 (same/different haplotype possessed by two core groups).

7.3.3: The Puzzle of Cooperation: Reciprocity. Reciprocity is based on the concept that the rate at which an individual interacts with another individual depends on the previous history of interactions in the community. Three principal forms of reciprocity have been studied, each theoretically leading to the evolution of cooperative behavior (Axelrod & Hamilton 1981; Mohtashemi & Mui 2003; Pfeiffer et al. 2005). Each makes predictions about the pattern of interactions in a population, and so the different forms of reciprocity can be set up as tests on an interaction matrix:

- *Direct reciprocity.* This is the original, dyadic form of reciprocity, in which individuals direct more affiliative interactions or fewer agonistic ones toward other individuals who are more affiliative or less agonistic to them (Trivers 1971). As discussed by Hemelrijk (1990b; see also Section 4.8), the actor-receiver version of direct reciprocity is both more parsimonious and tractable. In this scheme, an individual interacts with other members of the population at rates that are proportional to the rates at which the other individuals interact with it. Hemelrijk (1990b) suggests using the R_r test to test for evidence of direct reciprocity. In this, the actor-receiver matrix and its transpose, the receiver-actor matrix, are first ranked within rows and then subjected to the Mantel test (e.g., Table 4.10).

· *Reputation reciprocity.* In reputation reciprocity, individuals tend to interact positively with those who interact positively with others. In other words, reputation counts. For an affiliative interaction measure, a test of this is that the interaction rate between A (actor) and B (receiver) is proportional to the overall (summed over the population) rate at which B acts. Once again, we may want to use the R_r version of the test, ranking within rows both the actual interaction matrix and expected interaction matrix, before carrying out the Mantel test.

· *Generalized reciprocity.* Generalized reciprocity, in which individuals who have been the subject of positive interactions act more positively, is the least cognitively demanding of the forms of reciprocity—an individual does not need to recognize other individuals or track their actions, just respond to the way it has been treated. It would also appear to be the least likely to evolve. One might wonder why an individual would not just receive and not give if no one is keeping track. Pfeiffer et al. (2005) show theoretically that generalized reciprocity can evolve, at least in small groups. The prediction of generalized reciprocity is, in some respects, the transpose of that from indirect reciprocity: that the interaction rate between A (actor) and B (receiver) is proportional to the overall (summed over the population) rate at which A receives. In this case, we might want to rank both the interaction matrix and expected interaction matrix within columns.

Table 7.2 illustrates these tests (the ranked versions) using the capuchin grooming data of Table 4.4. This data set provides little support for any of the forms of reciprocity, although direct reciprocity has greatest support (the highest matrix correlation coefficient).

Reputation reciprocity and generalized reciprocity could be applicable to a symmetric interaction measure. For instance, for an association measure based on grouping, reputation reciprocity would imply that individuals associate most with the most gregarious members of the population, and generalized reciprocity that more gregarious individuals in general are more likely to associate with particular partners. In both cases, however, the test would need to be modified to remove circularity (e.g., in reputation reciprocity, removing individual A when calculating B's gregariousness for the expected value of the A-B association). I am

Table 7.2 Tests for Reciprocity in Grooming Rates of Female Capuchin Monkeys (*Cebus capucinus*) Using Data of Perry (1996; table 4.4)

Observed: Expected with reciprocity:

Testing for direct reciprocity

Row-ranked matrix   Row-ranked transposed matrix

	A	S	N	D	W	T		A	S	N	D	W	T	
A	0	5	4	2	3	1	A	0	5	2	4	3	1	
S	5	0	4	3	2	1	S	5	0	5	3	4	1	
N	4	5	0	2	3	1	N	1	5	0	3	4	2	
D	5	2	3	0	4	1	D	1	4	3	0	5	2	$r = 0.35, P = 0.151$
W	4	3	2	5	0	1	W	1	3	4	5	0	2	
T	1	2	4	3	5	0	T	1	4	2	3	5	0	
	A	S	N	D	W	T		A	S	N	D	W	T	

Testing for reputation reciprocity

Row-ranked matrix   Row-ranked column-sums of matrix

	A	S	N	D	W	T		A	S	N	D	W	T	
A	0	5	4	2	3	1	A	0	5	2	3	4	1	
S	5	0	4	3	2	1	S	1	0	3	4	5	2	
N	4	5	0	2	3	1	N	1	5	0	3	4	2	
D	5	2	3	0	4	1	D	1	5	3	0	4	2	$r = 0.15, P = 0.369$
W	4	3	2	5	0	1	W	1	3	3	4	0	2	
T	1	2	4	3	5	0	T	1	5	2	3	4	0	
	A	S	N	D	W	T		A	S	N	D	W	T	

Testing for generalized reciprocity

Column-ranked matrix   Column-ranked row-sums of matrix

	A	S	N	D	W	T		A	S	N	D	W	T	
	A	S	N	D	W	T		A	S	N	D	W	T	
A	0	2	1	1	1	1	A	0	5	5	5	5	5	
S	5	0	5	4	3	4	S	4	0	4	3	3	3	
N	2	5	0	3	4	2	N	5	4	0	4	4	4	
D	4	3	3	0	5	3	D	3	3	3	0	2	2	$r = -0.17, P = 0.665$
W	3	4	4	5	0	5	W	2	2	2	2	0	1	
T	1	1	2	2	2	0	T	1	1	1	1	1	0	
	A	S	N	D	W	T		A	S	N	D	W	T	

In each of the three tests, the observed rankings of the grooming rates is shown on the left and the expected ranks on the right. The matrix correlation (*r*) and value of the Mantel test (*P*) is shown for each hypothesis.

not sure that such tests for reputation or generalized reciprocity using symmetric measures have a great deal of value.

7.3.4: Testing the Bond. Zahavi (1977) proposed that bonded animals routinely test the bond (Section 4.10), inflicting stress on each other and seeing how their partner responds. One possible method of testing a bond is by exchanging costly signals. Thus, we might expect a correlation across dyads between bond strength and the rate of costly signaling.

Whitham and Maestripieri (2003) examined this hypothesis using observations of 15 captive adult male baboons (*Papio papio*). They used a Mantel test to look at correlations between bond strength and rates of intense greeting. Alternative hypotheses were that the greetings were involved in aggressive or submissive behavior, and these also were translated into Mantel tests. The dyadic rate of intense greeting was correlated with grooming rate (Mantel $P < 0.0001$) and time spent associated (Mantel $P < 0.01$), although a partial Mantel test showed that this latter relationship no longer held when association rate was controlled for grooming rate ($P = 0.46$). Thus, intense greetings seemed related to the quality of the relationship—the bond strength. Arguing against dominance or submissive functions for intense greetings were a lack of significant relationships between intense greetings and rank difference (Mantel $P = 0.30$) or aggression rates (Mantel $P = 0.46$; partial Mantel $P = 0.22$ controlling for time spent in proximity). Rates of less intense, and so less costly, greetings were not significantly correlated with any of the affiliative or agonistic measures (Mantel $P > 0.05$), again in line with the bond-testing hypothesis. In this summary of Whitham and Maestripieri's (2003) results, I have just listed P-values, not a recommended practice ("naked P-values"; Johnson 1999). However, the effect sizes are presented as dyadic plots of interaction rates in the original paper (Whitham & Maestripieri 2003; e.g., Fig. 2).

BOX 7.1 *Quantitative Methods in the Behavioral Ecology of Social Systems*

The quantitative methods used by behavioral ecologists to investigate the function of behavior are usually closer to the statistical mainstream than those described in the earlier chapters of this book. They are mostly well covered by standard statistical texts (e.g., Sokal & Rohlf 1994). Here I highlight some of the most important methods (see also Table 2.1.) and issues, focusing first on experiments and then on comparisons of natural variation.

Experiments. The experimental method involves changing, randomly, the level of a factor that is hypothesized to affect some outcome. In the context of this chapter, social elements could be part of such analyses in three ways:

· Social measures as outcomes, potentially affected by nonsocial factors. For instance, the temperature or food availability of a captive colony could be altered on some random basis and social variables, either at the individual, dyadic, or societal level (Table 1.2), could be measured.

· Nonsocial outcomes, potentially affected by social factors. Manipulating social factors is difficult but sometimes possible, usually with captive colonies. An example of this kind of experiment might address how removal of an alpha male affects conception rates of females. Ideally, however, a number of separate colonies would need to be available and open to random removal of alpha males.

· Social measures as outcomes, potentially affected by social factors. Here, the difficulties are increased with less tractable social measures to be both manipulated and measured. A fine example of this type of approach, however, is Flack et al.'s (2006) comparison of network measures of social structure in a macaque colony from which three "policing" individuals were removed randomly.

The basic analytical method of analyzing such experiments is the analysis of variance (ANOVA) or t-test if there are only two levels of the factor. Variants include the multivariate analysis of variance (MANOVA) when two or more outcome measures are to be analyzed together.

Several issues should be considered when planning experiments and analyzing experimental data. These include the normality of the response variable(s). If the response variables are non-normal, it is best to transform them so that they are normal or nearly normal (Sokal & Rohlf 1994, pp. 409–422). In situations in which the response variables cannot be made normal, one can use the nonparametric equivalents of the ANOVA and t-test (Mann-Whitney U-test and Kruskal-Wallis test, respectively). If there are several response measures, then analysts might want to adjust the critical levels to control experiment-wise probabilities of rejections of the null hypothesis that there is no effect of the experimental manipulation. Bonferroni and Dunn-Sidak corrections do this (Sokal & Rohlf 1994, p. 240). Generally, it is even better to use the multivariate form of the test (MANOVA) so that only one P-value is generated despite there being several response variables.

The assumption of independence is one of the most likely problems to affect experimental studies of social behavior. Statistical methodology generally assumes random samples from the study population. This is rarely the case if we consider the study population to be the species in the wild or captivity. Thus, repeated measurements made on the same individual are not independent, nor are repeated measurements made on the same community, unless the community itself is considered to be the study population. Thus, for instance, the results of Flack et al.'s (2006) otherwise exemplary study of policing technically only refer to the macaque colony under observation during the period of the experiment. Extrapolation of the results more generally requires a leap of faith that is not statistically buttressed.

Comparison of individuals, dyads, populations or species. In these nonexperimental situations, variables are compared. The simplest case is of two continuous variables, perhaps a societal measure (e.g., from Table 1.2) and an environmental one, such as food availability. These can be related using the correlation coefficient between the measures (Section 2.2), which expresses how closely related they are. In addition, it is always worth plotting them against one another because a correlation coefficient might mask many kinds of interesting nonlinearities between the measures. The correlation coefficient can be, and usually is, tested against zero. If the data are quite nonnormal, then the Spearman rank correlation coefficient (Section 2.2) is a suitable alternative.

Regression analysis can be used to quantify the relationship between the variables, assuming that one can be considered dependent on the other. There are many variants of regression analysis, including multiple regression when there are several independent variables, polynomial regression for nonlinear situations, and logistic regression when the dependent variable is categorical (Kleinbaum et al. 1988). If it is the independent variable or variables that are categorical, then this turns into an ANOVA. If both dependent and independent variables are categorical, then we use log-linear models (Sokal & Rohlf 1994, pp. 743–760). All of these are special cases of the generalized linear model, which allows for combinations of categorical and continuous dependent and independent variables as well as nonlinear relationships between them (Dobson 2001).

With more than one dependent variable, the generalization of correlation analysis is to canonical correlation analysis (Legendre & Legendre 1998, pp. 612–616) and that of multiple regression is to redundancy analysis (Legendre & Legendre 1998, pp. 579–594).

From these methods, we obtain estimates of the importance of the relationships among the variables (e.g., correlation coefficient, proportion of variance accounted for), descriptions of the relationships (e.g., regression coefficients, effect sizes), and the results of tests of the null hypothesis that there is no relationship. These may all be useful, but many have argued that the results of tests should not be presented without measures of the size or nature of the relationship, and that confidence intervals about such measures are more revealing than P-values (e.g., Johnson 1999).

All of these techniques make assumptions. Most assume that residual errors, after fitting the regression or other model, are normally distributed with constant variance and that there is some kind of linear relationship between the two variables or sets of variables. Residuals are usually used to check these assumptions (Kleinbaum et al. 1988, pp. 185–196). If these assumptions fail, transformations of the variables, by logging, ranking, or in other ways, are usually the best way to proceed. Sometimes, a more complex method, such as a generalized linear model or even a nonlinear model, can be used to deal with assumption failure.

As with the experimental method, in comparative studies the assumption of independence is a particularly significant one. Dependences take rather different forms depending on the nature of the units of the analysis, whether they are individuals, dyads, populations, or species.

If the level of comparison is the individual, then either only one entry should be made for each individual in the analysis and each variable (although it could be the mean, or some other statistic, calculated from several measurements), or individual identity should be entered into the analysis as an independent factor. Even after one takes these steps, there may still be problems with independence if, for instance, some of the individuals are close relatives.

When the units of analysis are dyads, the independence problems become more serious because the same individual is responsible for a number of the units in the analysis. As discussed in the earlier chapters, however, there are ways around this. Measures of precision such as confidence intervals and standard errors can be estimated using bootstraps and jackknives (Section 2.3), whereas hypothesis tests are best achieved through permutation (Section 2.4). The Mantel test is particularly valuable in this situation (e.g., examples in Sections 4.8, 4.11, 7.3, and 7.6). We can also use the partial Mantel test (Smouse et al. 1986) to control for other aspects of nonindependence (e.g., same–different sex).

When populations or communities are the units of analysis, lack of independence is not built in, but it still could be an issue, and so we might need to use controlling covariates, such as population size or proportion of mature males, when, for instance, looking at how prey availability affects group size. In some cases, such as when we want to control for geographical distance between populations, it might be useful to recast the problem as a Mantel test (Section 2.4) so that, instead of comparing group size with prey availability, we compare difference in group size with difference in prey availability, controlling for geographic distance.

Finally, when species or other phylogenetic entities are the units of analysis, there is an additional problem with independence. Evolutionarily related species may have similar values on a trait either because of homology or because there has been parallel evolution of, or ecological adaptation to, the independent variable being considered (Stearns & Hoekstra 2000, p. 323). To get around this, scientists use the method of independent contrasts, in which comparisons are made between pairs of species, or groups of species, such that each divergence in the phylogenetic tree is only used for one comparison (Felsenstein 1985). The method of independent contrasts is often regarded as the optimal technique when comparing species (Stearns & Hoekstra 2000, pp. 327–328). However, the method depends on an accurate phylogenetic tree being available, usually from molecular genetics, and if the tree is inaccurate, the analysis may become problematic. The program CAIC implements the method of independent contrasts (Purvis & Rambaut 1995).

7.4 *Environmental Determinants of Social Systems*

In this section and the next, which considers the relationship between social and mating systems, we reach ground that has been much traveled. Among the many important contributions on the evolution of social and mating systems are those of Wilson (1975), Emlen and Oring (1977), Clutton-Brock (1989), and Kappeler and van Schaik (2002). Many other papers and books could have been chosen, but this selection indicates both a range of dates and taxonomic perspectives, with the authors working primarily on insects, birds, nonprimate mammals, and primates. There is much that is unknown and contested in this area, but some general principles have emerged. Here and in the following section, I summarize the principles—for more details, there is plenty to read!—and discuss methods. See also Box 7.1.

Organisms adapt to their environments, and, in this respect, social relationships can be treated like any other trait. However, there is major complicating factor: Because social behavior involves two or more organisms, each of which is part of the others' environment, the system is potentially challenging both for the evolutionary optimization of an organism's behavior and its study. Despite these challenges, it has become clear that nonsocial factors can be major determinants of social structure in some simpler systems, as well as providing an overall context for the evolution of more complex ones. Aspects of the nonsocial environment that are considered to be major determinants of social structure include the following:

- *Predation.* The presence of nearby conspecifics can help combat predators through increased vigilance, dilution (the predator is more likely to take someone else), confusion of the predator (e.g., synchronous movement of shoaling fish), cover (hiding behind group members), and group defense, for instance, "mobbing" in birds. Thus, predation is often considered to be a force for gregariousness (Pulliam & Caraco 1984). However, there may be drawbacks to forming groups. For instance, grouping may make prey more conspicuous or more desirable, as when the predator is a bulk feeder, such as a baleen whale that only attacks shoals of fish and not solitary individuals.

- *Finding food.* Animals can use other animals to help them find food in a variety of ways, including information centers (in which, for instance, roosting birds note the incoming bearing of conspecifics who have fed successfully and leave the roost on its reciprocal bearing), local enhancement (reacting to the feeding behavior of others), flushing prey, or communally searching large areas for widely dispersed prey (Pulliam & Caraco 1984). In some cases, however, living in groups can make finding food harder, for instance, if the prey are easily scared into safe refuges or there is a scramble competition, with groupmates eating the potential food of an individual.

· *Catching food.* Sometimes catching prey is easier with cooperation, and in particular carnivores, such as lions (*Panthera leo*), can bring down larger prey when working together.

· *Defending resources.* Grouping may be adaptive in that the members can defend food or other resources against members of their own or other species (Wrangham 1980).

· *Patchiness of resources.* If resources are patchy, this will tend to aggregate animals. Although such aggregations are not social in the way in which I use the term (Section 3.4), aggregation can act as a catalyst for sociality; animals must be within communicative range of one another to interact, and patches of resource may bring them within such range.

The significance of these and other hypotheses in determining social structure is primarily examined through comparisons of species, populations of the same species in different environments, or changes in the social structure of a particular population as environmental conditions change (Box 7.1). Such comparisons are often qualitative, at least from the perspective of characterizing the social structure. However, measures of social structure (Section 5.1) and methods of characterizing it (e.g., Section 5.2–5.5) can add rigor or deeper insight. Here are some examples.

Hill and Lee (1998) compared community sizes (which they, like other primatologists, call "group sizes") and predation rates for 121 populations of 39 primate species. Mean community sizes were almost twice as high ($\sim$70 animals) in populations with a high risk of predation ("frequent predator–prey interactions with regular contact and actual or attempted predation observed or suspected") than in populations with a medium ($\sim$28 animals) or low ($\sim$21 animals) risk. There were unexpectedly high numbers of mature males in populations with high predation risk.

As with other correlative studies, there is the possibility that some other factor, such as phylogeny, body size, or habitat productivity, might have affected both community sizes and predation risk, producing a relationship between the two factors without any direct causative mechanism. Analytical techniques such as multiple regression and the method of independent contrasts (e.g., Stearns & Hoekstra 2000, pp. 323–326) can help to distinguish such explanations (Box 7.1).

Faulkes et al. (1997) used independent contrasts to look at ecological influences on the evolution of sociality in African mole rats. Mole rats are subterranean rodents. For mammals, they have a uniquely wide variety of social structures, ranging from solitary to eusocial (characterized by cooperative care of the young and sterile castes). Figure 7.6 shows a phylogeny determined by molecular genetics together with group sizes and type of social structure for each species or subspecies when known. It is clear that group sizes and social structures are often similar among closely related species and subspecies, and thus that the species and subspecies are not independent. Therefore, Faulkes et al. (1997) made independent contrasts at the nodes marked A through G on Fig. 7.6. At each node, they compared the difference in social variables with the difference in environmental variables. Here is the essence of the results of correlation analysis using these contrasts with the logarithm of the maximum group size being correlated with several ecological and other independent variables (Faulkes et al. 1997):

Log mean geophyte density	$r = -0.778$ $(P = 0.022)$
Log mean digestible energy	$r = -0.527$ $(P = 0.179)$
Log mean annual rainfall	$r = 0.003$ $(P = 0.993)$
Months of >25-mm rainfall	$r = -0.223$ $(P = 0.595)$
CV of rainfall	$r = 0.915$ $(P = 0.001)$
Mean body mass	$r = -0.469$ $(P = 0.241)$
Gestation length	$r = 0.522$ $(P = 0.229)$

This indicates that social systems in mole rats are related to geophyte density and variation in rainfall, with the more complex social systems being found in habitats of low geophyte density and high variability in rainfall. Thus, in this group of species, complex social systems seem to have arisen in more challenging environmental conditions.

For an example of the effects of environment on social structure that includes comparisons within and between populations, consider the study of Henzi et al. (Submitted). They used lagged association rates (Section 5.5) and network analysis (Section 5.3) to study the effects of seasonal environmental differences on the societies of female chacma baboons (*Papio hamadryas ursinus*). When food was abundant, models fitted to lagged association rates included components indicating preferred companions, whereas when food was scarce, simpler models without such terms fitted better. In addition, when food was abundant, network strengths (i.e., gregariousness) and clustering coefficients ("cliqueishness") decreased, but the variability in the association indices of individuals increased (in agreement with the lagged associa-

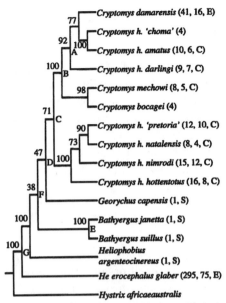

FIGURE 7.6 Phylogeny of African mole rats (family Bathyergidae) determined using molecular genetics together with maximum colony size, mean colony size, and type of social system when known. C, colonial, with usually one breeding pair per colony; E, eusocial; S, solitary. Beside nodes are shown the percentage of bootstrap replicate phylogenies containing the node. (From Faulkes et al. 1997, fig. 1a.)

tion rate analysis). These patterns were consistent at two different sites (even though the seasons of food abundance are different) and over two changes in season at each site.

7.4.1: Are Social Systems Adaptive? The preceding examples show social structures varying with environmental conditions within and between species. Can we then call them adaptive? A state of a trait is adaptive if, compared with alternative states, it has improved reproductive performance (basically, surviving offspring). Darwinian evolution through natural selection predicts that traits will usually be adaptive (Stearns & Hoekstra 2000, pp. 13–18). This means adaptive at the level of the replicator, the unit of information and usually the gene, not necessarily adaptive at higher levels. In many cases, the actions of individual organisms will be adaptive, but sometimes "selfish" genes will make this to be not the case (Dawkins 1976). As we move up levels of organization away from that of the replicator, the likelihood of adaptation decreases. From this perspective, the social structure of a population would not necessarily be expected to be adaptive in the sense that the mean reproductive performance of its members might be higher with an alternative social structure. Examples of important elements of social structure that are nonadaptive from the perspective of the mean reproductive performance of its members are the presence of infanticide (e.g., van Schaik & Kappeler 1997) and the formation of groups of size larger than the size that optimizes reproductive success (Sibly 1983). Such elements arise and persist because the individual behavioral decisions that cause them are adaptive to the individuals within their social environment, but the population as a whole would do better without them.

For social systems to be generally adaptive, communities would need to be the replicators themselves. This is "group selection," the conditions for which are believed to have little likelihood of occurring (Stearns and Hoekstra 2000, p. 50), except in the case of cultural group selection, in which the transmission of traits is through social learning rather than genes (Soltis et al. 1995).

A second line of argument about the general adaptiveness of social systems comes from studies of self-organization. As reviewed by Sumpter (2006), a variety of empirical and theoretical studies of a diverse range of species have shown that complex social behavior can arise from the rather simple actions of individuals. In his abstract, but nowhere else in the article, Sumpter (2006) uses the word "adaptive" to describe the complex social patterns that may result. The collective behavior is, I believe, likely usually individually adaptive in the sense that if it is prevalent, an individual does best, in terms of reproductive performance, by joining in, but not necessarily collectively adaptive because there may be some other form of social structure that might have increased mean reproductive performance.

7.5 Mating Systems and Social Systems

The mating system, as I have defined it, is a part of the social system. In most cases, mating and its related activities such as courtship, copulation, mate guarding, and mate competition, form a large and important part of the set of interactions and relationships that define a social system. Mating is also behind many other significant interactions and relationships, including infanticide and alliance formation.

The relationships between mating systems in particular and social systems in general have long been a focus for the attention of behavioral ecologists (e.g., Wilson 1975) and much has been written about them. In some groups of species, such as pinnipeds and amphibians, most research on social systems has actually been about mating systems. The relationship can work in both directions. For instance, the group sizes of females are considered to be an important element in the type of mating system adopted by mammals (Clutton-Brock 1989), whereas infanticide, a male mating strategy, can be a force for general female sociality (Ebensberger 1998).

Important elements in the evolution of mating systems include the following (Emlen & Oring 1977; Clutton-Brock 1989; Kappeler & van Schaik 2002):

- *Parental care*. How many adults are needed to care for the offspring? If it is two (as in many birds), then monogamous systems are likely to prevail; if it is one (as in many mammals), forms of polygyny, polyandry, or promiscuity are expected; if it is more than two (found occasionally among mammals and birds, but most prominently in the social insects), then complex helper-based societies evolve; if it is zero (some fish), then parental care is not an issue and other factors come into play.

- *Monopolization of mating opportunities*. When the mating opportunities of one or both parents are unconstrained by parental care, then their ability to monopolize mating opportunities with members of the opposite sex becomes important. This ability depends on the spatial distribution of the members of the opposite sex, and so on their social structure, as well as on their use of defendable resources and the temporal pattern of receptivity for mating. Polygyny may be expected if a single male or coalition of males can defend economically a group of receptive females, the resources on which they depend, or a sequence of receptive females (Emlen & Oring 1977).

- *Partner choice*. Females, investing much more in each gamete, are expected to be choosier. Partner choice is clearly a major part of many monogamous mating systems with elaborate courtship rituals, as well as in polygynous or promiscuous systems in which one sex (usually males) displays to the other (Trivers 1985, pp. 331–360).

- *Infanticide*. Parents may adopt tactics, including social behavior, to lessen the likelihood that their offspring are killed. Usually the threat comes from males interested in mating with the mother, who, with her offspring's death, may

become receptive more quickly. It has been proposed that alliances among females and promiscuity may be adaptations
against infanticide (Ebensberger 1998).

To study mating systems, we first need to determine their form. This
is done in three main ways: (1) from observation—who mates with
whom; (2) from genetic analyses—who were the parents of whom, and
thus mated; and (3) using inferences from other areas of biology. An example of the latter is using the correlation between relative testis size and
mating systems in primates, with males having relatively larger testes in
promiscuous systems (Harcourt et al. 1981), to predict mating systems
in baleen whales (Brownell & Ralls 1986), for whom observations of
mating are very rare, but there is plenty of data on testis size collected
by the whaling industry. Frequently, mating systems are described using
categorical variables, such as "monogamous," "polygynous," "polyan-
drous," and "rapid multiple clutch polygamy" (Emlen & Oring 1977).
Numerical variables might also be useful, however, such as the number
of males per social unit or the mean number of males with which each
female mates during a receptive period.

Because a mating system is the property of a population, or at least a
community, a quantitative analysis usually needs to have statistical units
of at least this scale. Experiments become difficult here. Mating systems
themselves cannot really be manipulated directly in any consistent but
randomized way. Especially in captive situations, it may be possible to
manipulate the social systems behind the mating systems, for instance,
by changing the sex ratio in a colony, but this is still demanding (Box
7.1). When mating systems are the response variable in an experimental design, they do not need to be manipulated, but because it usually
takes some time for a mating system to develop and be registered, such
experiments are also demanding. Consequently, mating systems are usually examined using intraspecific (e.g., Moehlman 1998) or, especially,
interspecific comparisons. Box 7.1 notes major methods.

As an example of the interspecific method, consider the study of Hasselquist and Sherman (2001). There are two principal hypotheses for how
the prevalence of extrapair copulations in birds should relate to the mating system. In the first, extrapair copulations should be higher in polygynous systems because males cannot as effectively protect females with
whom they have mated from being mated by other males as they could
in monogamous systems. Alternatively, if females are driving extrapair copulations, there might be more in monogamous systems because
they would be less likely to pair with the male of their choice. Thus,

Hasselquist and Sherman (2001) examined how the rate of extrapair copulations varied between mating systems for 40 species of passerine bird. Mating systems were classified as monogamous (<5% of males paired with more than one female) or polygynous (>5% of males pairing multiply). There were proportionally more extrapair young in monogamous species (mean 23%) than in polygynous species (mean 11%). Because of the phylogenetic relationships among the species, they used the method of independent contrasts (Fig. 7.7; Box 7.1) to test whether this difference was unexpectedly large. It was: There were significantly more extrapair young in monogamous species, whether mating systems were categorized as monogamous or polygynous ($P < 0.001$) or placed on a continuum measured by the proportion of polygynous males ($P = 0.02$). Thus, Hasselquist and Sherman's (2001) study strongly supports the hypothesis that it is the female's desire to mate with the "best" male that drives extrapair copulations, and that in polygynously mated species there is less drive for this and/or that females incur higher costs for sexual unfaithfulness.

7.6 Culture in Society

7.6.1: What Is Culture? Culture has scores of definitions (Mundinger 1980), but most of them have the same basic elements as the one that I use: "information or behavior shared by a population or subpopulation of animals that is acquired from conspecifics through some form of social learning" (Rendell & Whitehead 2001). Culture thus constitutes an additional way that behavior is transmitted from animal to animal— additional to genes, that is. Social learning is very different from reproduction: Most vertebrates receive genetic input from just two parents at conception, whereas social learning can occur throughout life and from many models. Individuals may easily modify the information that they learn socially before using it or passing it on, whereas this rarely occurs with genetic information. Thus, in several respects, cultural evolution operates quite differently from genetic evolution, and its products tend to have different characteristics (Richerson & Boyd 2004). These include a greater likelihood of apparently maladaptive behavior, more possibilities for group selection, and conformist, culturally marked groups. All of these are attributes of the "hypercultural" (Barkow 2001) societies of modern humans (Richerson & Boyd 2004). Perhaps other species "have culture." If so, how do we recognize it and investigate its importance?

There are two principal ways in which nonhuman culture is studied, focusing on two of its fundamental attributes: the "shared" property

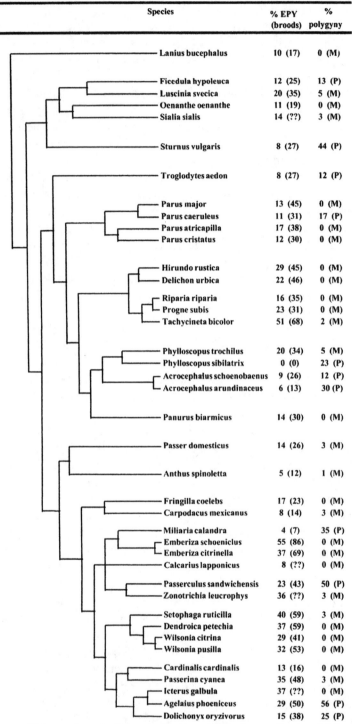

Species	% EPY (broods)	% polygyny
Lanius bucephalus	10 (17)	0 (M)
Ficedula hypoleuca	12 (25)	13 (P)
Luscinia svecica	20 (35)	5 (M)
Oenanthe oenanthe	11 (19)	0 (M)
Sialia sialis	14 (??)	3 (M)
Sturnus vulgaris	8 (27)	44 (P)
Troglodytes aedon	8 (27)	12 (P)
Parus major	13 (45)	0 (M)
Parus caeruleus	11 (31)	17 (P)
Parus atricapilla	17 (38)	0 (M)
Parus cristatus	12 (30)	0 (M)
Hirundo rustica	29 (45)	0 (M)
Delichon urbica	22 (46)	0 (M)
Riparia riparia	16 (35)	0 (M)
Progne subis	23 (31)	0 (M)
Tachycineta bicolor	51 (68)	2 (M)
Phylloscopus trochilus	20 (34)	5 (M)
Phylloscopus sibilatrix	0 (0)	23 (P)
Acrocephalus schoenobaenus	9 (26)	12 (P)
Acrocephalus arundinaceus	6 (13)	30 (P)
Panurus biarmicus	14 (30)	0 (M)
Passer domesticus	14 (26)	3 (M)
Anthus spinoletta	5 (12)	1 (M)
Fringilla coelebs	17 (23)	0 (M)
Carpodacus mexicanus	8 (14)	3 (M)
Miliaria calandra	4 (7)	35 (P)
Emberiza schoeniclus	55 (86)	0 (M)
Emberiza citrinella	37 (69)	0 (M)
Calcarius lapponicus	8 (??)	0 (M)
Passerculus sandwichensis	23 (43)	50 (P)
Zonotrichia leucrophys	36 (??)	3 (M)
Setophaga ruticilla	40 (59)	3 (M)
Dendroica petechia	37 (59)	0 (M)
Wilsonia citrina	29 (41)	0 (M)
Wilsonia pusilla	32 (53)	0 (M)
Cardinalis cardinalis	13 (16)	0 (M)
Passerina cyanea	35 (48)	3 (M)
Icterus galbula	37 (??)	0 (M)
Agelaius phoeniceus	29 (50)	56 (P)
Dolichonyx oryzivorus	15 (38)	25 (P)

FIGURE 7.7 Phylogeny of passerine bird species showing percentage of young with extra-pair fathers (EPY; percentage of broods with more than one extrapair father) and percentage of males paired with more than one female. M, monogamous (< 5% of males have more than one social mate); P, polygynous (> 5% of males have more than one social mate). (From Hasselquist & Sherman 2001, fig. 1.)

and "social learning" (Laland & Hoppitt 2003). Social learning comes in several forms, including imitation, emulation, teaching, and social enhancement (Whiten & Ham 1992). It is usually studied experimentally or through detailed observational studies of captive animals (Laland & Hoppitt 2003). Unfortunately, the species for which this approach is feasible are not always those whose behavior seems to have the most cultural input, which is certainly the case for the sperm whales (*Physeter macrocephalus*) that I study. Even when the experimental study of social learning is available, the experimental laboratory situation is invariably so different from the real-world environment of the animals that the utility of the results for the assessment of culture in the wild is questionable (McGrew 1992, p. 21; Boesch 2001).

Thus, field biologists interested in culture usually focus on another of its attributes: that it is a group phenomenon. Different sets of animals can have different cultures. Using this approach, we try to identify these sets of animals and define their distinctive behavior. A major challenge is determining that these behavioral distinctions result from social learning rather than the principal alternatives of genetic determination, individual learning in different environmental circumstances, or ontogenetic (age-related) effects. How do we proceed? This is an area in development (Laland & Janik 2006). In the following subsections, I consider the identification of social entities for cultural analysis, the current general protocol, which features the elimination of noncultural causes of observed behavioral patterns, and a new approach, which attempts to partition the variance in behavior into cultural, genetic, environmental, and possibly other causes.

7.6.2: Identifying Cultural Entities. In this step, we look for sets of animals with similar behavior. Ideally, the sets are determined based on the feasibility of members learning socially from one another. Social learning itself is rarely unequivocally observed in the wild (but see, e.g., Boesch & Boesch-Achermann 2000, pp. 243–246). Although social learning between two animals is not the same as a behavioral interaction, the circumstances under which they occur are likely similar. Thus, it makes sense to define association in a way similar to that used when measuring relationships (Section 3.3), construct association indices that indicate the probability that social learning could take place within a dyad (Section 4.5), and divide the population into sets (units, communities, or other social entities) based on the pattern within the matrix of association indices (Section 5.7). This could include two or more hierarchically organized tiers of social structure or two or more non–hierarchically arranged social patterns (Section 5.7).

In some situations, defining suitable social entities is simple. For instance, chimpanzee (*Pan troglodytes*) communities are well defined by defended territorial boundaries. In other cases, such as sperm whales (Whitehead 2003), social units need to be delineated by analytical processes such as those considered in Section 5.7.

For the remainder of this section, I call these sets of animals "communities" because they usually fit within my definition of community, being largely behaviorally self-contained and within which most individuals interact with most others. However, this need not be the case. For instance, with humpback whales (*Megaptera novaeangliae*), a usual division when looking at the cultural properties of songs is an ocean basin (Payne 1999), but most pairs of individuals within an ocean basin do not interact.

Occasionally (e.g., Krützen et al. 2005) the cultural communities are defined based on the behavioral activity being considered. Thus, animals doing one thing are allocated to one community and those doing another to a second community. This procedure makes the rigorous identification of culture, and especially the ruling out of environmental influence, harder. Even if there seems to be no difference in the environments of the animals behaving differently, how do we know that some unmeasured environmental variation did not cause the difference in behavior? Because the communities are defined using the pattern of behavior itself, environmental influences can much more easily cause culture-like patterns than when the communities are delineated in an independent study based on associations that indicate the probability of social learning.

7.6.3: The Elimination Method: Is It Culture? Once the communities have been identified, a set of candidate behavioral activities can be mapped onto them. These activities could be vocalizations, visual or tactile signals, foraging methods, food types, movement characteristics, types of play, social conventions, or other types of behavior (for examples, see Whitehead 1999b; Deecke et al. 2000; Whiten et al. 2001; Mann & Sargeant 2003; Perry & Manson 2003; van Schaik et al. 2003). An activity is a candidate for culture if it is found habitually (occurring repeatedly in several individuals within the community) (Whiten et al. 1999) in some of the communities but not in others, or, if it is a quantitative measure, at different rates or intensities in the different communities. The "elimination" protocol for identifying culture next operates by ruling out alternative explanations for the patterns identified. The alternatives are as follows.

1. *Ontogeny*. This is usually the easiest to eliminate if all the communities contain individuals of an age at which this activity is performed. This step is often omitted if the sets are virtually closed breeding units.

2. *Environmental variation*. The suggestion here is that the pattern of occurrence of the behavior could have arisen because individuals in the different communities are exposed to different stimuli and thus learn, individually, different behavior, or the environment induces behavioral differences through some other form of phenotypic plasticity. In some cases, this alternative is easily ruled out, for instance, when the communities are sympatric using the same environment, or especially with some communicative or social behavior, when the asocial environment can reasonably be considered to have no relevance. In many other cases, especially with foraging activities, it is much harder to rule out environmental determination, and "elimination" studies of culture are often criticized on the basis that this has not been done sufficiently carefully (Laland & Janik 2006).

3. *Genetic determination*. Genetic determination can be ruled out in a number of ways. Cross-fostering experiments are perhaps ideal, in which young animals are moved between communities with different behavior. If the fostered individuals show the behavior of their adoptive parents and social partners rather than that of their biological parents, then genetic determination can be ruled out. The situation in which immigrants adopt the behavior of the community that they enter is nearly as persuasive. If community-specific behavior changes with time (but not age), so that the members of a community adopt a form of behavior and then change it gradually or suddenly (as with humpback whale songs; Payne 1999) over periods of time less than a lifetime, then genetic determination can be ruled out. Often, none of these kinds of evidence is available, and we fall back on molecular genetic data. In such cases, the pattern of use of the activity within the communities becomes important. If virtually all members of some communities do use the activity and virtually all members of the other sets do not (often the case with discrete social signals, such as vocalizations), then genetic determination can virtually be ruled out if the communities with the trait are not monophyletic with respect to

the activity (which would imply that all individuals with the activity are descended from a common ancestor who is not the ancestor of any individual without the activity). This is close to the procedure used by Whiten et al. (1999) in their analysis of the cultural activities of chimpanzees, in which they showed that variation in patterns of activity was as great among communities within two subspecies as between subspecies. If the variation in the activity among communities is quantitative or not shared among nearly all members of a community when it is present, then ruling out genetic determination is harder, and methods are not properly worked out. A Mantel test (Section 2.4) that compares genetic and behavioral similarities among communities is one approach, with a near-zero matrix correlation and nonsignificant test result indicating that genetic determination is unlikely.

A difficulty here is that if communities are formed by fission and the daughter communities retain the cultural attributes of the parent community, then genetic and behavioral measures will tend to concur, giving a positive matrix correlation and a significant result for the Mantel test, even though the behavior may be culturally determined. If the animals mate within communities, then disentangling the two effects is extremely hard. If mating is usually between communities or if one sex usually migrates between communities before mating, then there is a way around this. Take the case in which communities are largely matrilineal, with females generally mating with males born in different communities. There may be a correlation among communities between mtDNA genetic similarity and behavioral similarity, but if there is no correlation between measures of biparental autosomal genetic similarity (e.g., using microsatellites) and behavioral similarity, then culture is probably the most parsimonious cause of the behavioral variation among communities because behavior is generally expected to be determined by autosomal genes [although there are some exceptions; for a counterargument see Janik (2001)].

7.6.4: The Regression Method: How Important Is Culture? The standard elimination protocol described in the previous subsection is restrictive; if a behavioral activity survives the protocol's rigorous application, it is likely to be culturally determined. However, culturally determined activities may be present in all studied communities or be correlated with environmental variation (especially foraging activities) or genetic variation (especially when most mating is within communities). In all these situations, the "culture-as-last-resort" elimination protocol just outlined would conclude "no evidence for culture." There is no reason that I find valid why culture in nonhumans should be invoked only when all other explanations have failed, especially when studies of human behavior take the opposite approach (Laland & Hoppitt 2003). We need a method of partitioning the causes of behavior into genes, environment, and culture (Laland & Janik 2006). Here is a potential way forward that I recently proposed (Whitehead In press-b), which uses multiple regression on association matrices (Legendre & Legendre 1998, p. 559).

Each of the matrices is indexed by the individuals in the population. Consider four such matrices:

- *Behavioral similarity.* This is the dependent variable— differences among individuals in the behavioral trait whose cultural determination we wish to examine. It could be binary (1:0), for instance, whether the individuals use the same type of greeting vocalization or do not. Perhaps more often, it will be continuous. When studying foraging behavior, diet similarity could be measured by a niche overlap index (Krebs 1989, pp. 379–380), whereas if the focus is vocal behavior, dialect similarity could be measured by the proportion of call types shared (Yurk et al. 2002).
- *Genetic relatedness.* This indicates the probabilities of the individuals sharing a gene through common descent and thus acquiring the same behavior genetically (e.g., Table 4.2). Estimates of genetic relatedness are normally the product of molecular genetics, nowadays often produced using microsatellites, but they could be derived from knowledge of genealogy (Section 4.2).
- *Ecological similarity.* This measures the similarity of the habitats of each pair of individuals. Within a single study area, this might be indicated by proportional range overlap using data from tracking studies. When comparing several study areas, ecological similarity measures (Krebs 1989, pp. 293–309) might be appropriate.
- *Social similarity.* In the context of culture, this is the key measure, indicating the probability that two individuals learned the same behavior socially. As suggested in the previous subsection, we can use association indices (Section 4.5) for this, but we need to think carefully about the

definition of association. It should be defined in such a way that associated animals could learn from one another. For instance, if animals learn most from adult females and before their sexual maturity, then the association index might be restricted to data collected when the individuals were both in appropriate age/sex classes and zero association assigned if they never were in the characteristic learner–model classes at the same time. A standard association index should often suffice, however, so that for social similarity, we use something like the matrix of association indices in Table 2.5.

I suggest making a multiple regression of the nondiagonal elements of the behavioral similarity on the corresponding elements of the other matrices. Thus, for any dyad we have (Whitehead In press-b)

Behavioral similarity = constant + genetic relatedness

+ ecological similarity + social similarity + error

Now, we need to use the data to evaluate the importance of the four independent variables on the right-hand side of the regression equation. I suggest doing this using standard partial regression coefficients (Sokal & Rohlf 1994, p. 614). Standard partial regression coefficients range from −1 (perfect inverse linear relationship) through 0 (no relationship) to +1 (perfect linear relationship) and indicate the significance of genetic variation, ecology, and social learning in determining behavior, in each case controlling for the other independent variables. The jackknife (Section 2.3), in which individuals are omitted from the analysis in turn, can be used to estimate the precision of the standard partial regression coefficients (Whitehead In press-b).

The hypothesis that social similarity does not affect behavioral similarity can be tested using a permutation test (Section 2.4) in which individual identities in the social similarity matrix (often association indices) are randomly permuted to give a distribution of randomized standard partial regression coefficients with which the real standard partial regression coefficient can be compared (Whitehead In press-b). The significance of genetic relatedness or ecological similarity could be tested similarly.

This technique has not been used on real data. To illustrate its use, however, Table 7.3 shows the four similarity matrices using artificially constructed data [for another example using artificial data, see Whitehead (In press-b)]. In this case, the standard partial regression coefficients, jackknife standard errors, and their statistical significance using

Table 7.3 Simulated Similarity Matrices of Eight Individuals in Behavior (e.g., Proportion of Call Types Shared), Genetic Relatedness as Might Result from Molecular Genetic Analysis, Ecology (e.g., Proportion of Range Overlap)), and Social Similarity as from Association Indices

Behavioral similarity

	B	C	D	E	F	G	H
A	0.88	0.51	0.71	0.40	0.35	0.00	0.45
B		0.72	0.27	0.19	0.53	0.20	0.04
C			0.03	0.19	0.24	0.22	0.22
D				1.00	0.47	0.52	0.74
E					0.58	0.62	0.80
F						0.17	0.49
G							0.69

Genetic relatedness [a]

	B	C	D	E	F	G	H
A	0.29	0.01	0.09	0.12	0.51	0.10	0.50
B		0.36	-0.07	-0.01	0.17	-0.03	0.19
C			0.47	0.00	-0.19	-0.04	-0.02
D				0.10	0.45	0	0.24
E					0	-0.01	-0.04
F						0.10	0.52
G							-0.02

Ecological similarity

	B	C	D	E	F	G	H
A	0.91	0.84	0.51	0.17	0.04	0.08	0.20
B		0.55	0.37	0.33	0.10	0.02	0.16
C			0.26	0.25	0.27	0.05	0.14
D				0.52	0.82	0.53	0.21
E					0.58	0.75	0.33
F						0.78	0.40
G							0.81

Social similarity

	B	C	D	E	F	G	H
A	0.68	0.59	0.28	0.17	0.81	0.14	0
B		0.67	0.16	0.10	0.71	0.10	0.13
C			0.09	0.25	0.69	0.37	0.21
D				0.91	0.17	0.78	0.51
E					0.39	0.59	0.95
F						0	0.17
G							0.66

The elements of these matrices are input into a multiple regression so that the influences of genetics, ecology, and social affinity in determining behavior can be estimated.

[a] Although relatedness values are theoretically positive, values calculated from molecular data using popular techniques may contain small negative values.

the permutation method outlined earlier, are as follows for the three independent variables:

$$\text{Genetic relatedness} = 0.185 \ (\text{SE} = 0.177; P = 0.095)$$

$$\text{Ecological similarity} = 0.373 \ (\text{SE} = 0.231; P = 0.004)$$

$$\text{Social similarity} = 0.627 \ (\text{SE} = 0.226; P < 0.001)$$

Thus, behavioral similarities seem to be most determined by social similarity, but with some contribution from ecology and perhaps a little from genetics. This indicates that the behavior whose similarities are indicated in the top left part of Table 7.3 is substantially determined by social learning.

According to the definition of culture that I use, the behavior also needs to be communal as well as determined by social learning. The social similarity matrix in the bottom right part of Table 7.3 indicates two communities, {A, B, C, F} and {D, E, G, H} (confirmed using average linkage cluster analysis with modularity equal to 0.38). Using this division, we find that the behavioral similarities are substantially and significantly greater within (mean behavioral similarity = 0.64) than between (mean behavioral similarity = 0.29) the communities (Mantel test $P = 0.03$). Thus, the data of Table 7.3 indicate community-specific behavior produced by social learning: culture.

This method could be modified to suit different situations. If ecological factors are unlikely to influence some behavior, such as perhaps communicative behavior, then the ecological similarity matrix might be omitted. Conversely, other independent variables might be added. Examples include gender similarity, age difference (thus considering ontogenetic reasons for behavioral similarity), or more than one measure of ecological similarity (e.g., range overlap plus another measure of habitat similarity) or social similarity (based on two or more definitions of association or interaction) (Whitehead In press-b).

Another modification is to carry out an analysis of the relative importance of different causes of behavioral similarity at the level of the social unit or some other social entity. Behavioral similarity, ecological similarity, and genetic relatedness between units may then be the differences between the mean values for members of each unit or (especially in the case of genetic relatedness) the mean of differences between dyads with one member in each unit. We need a different measure of social similarity, however, because units will not usually learn directly from one another.

Units between which there were recent transfers, or which through the processes of unit fission possessed a common ancestral unit, will tend to behave similarly if the behavior is culturally determined. Ancestral similarity in matrilineal units can be indicated by mitochondrial DNA. We have used mitochondrial DNA as a marker of ancestral links among social units in sperm whales (Whitehead et al. 1998). In this and similar (e.g., Krützen et al. 2005) cases, mitochondrial haplotype similarity is used to indicate social similarity, and the assumption is made that mitochondrial DNA does not code for the behavior being examined.

The method for partitioning behavioral similarity into genetic, ecological, and social causes makes several assumptions (Whitehead In press-b), including that the measurements are without error. Errors, which are perhaps most likely in the measures of ecological similarity, tend to reduce estimated standard partial regression coefficients from their true values, and so, in this sense, the method is conservative. Another potential problem with this, or any other, regression method is collinearity (Kleinbaum et al. 1988, pp. 206–218). If two or more of the independent variables (genetic, ecological, or social similarity) are closely linearly related then the regression analysis cannot partition their effects. For instance, if there is little preference for social partners, ecological similarity based on range overlap might be highly correlated with social similarity from an association index. In such cases, the failure of the regression technique due to collinearity indicates a genuine inability to separate the effects of the independent variables using the data and thus, in this case, to distinguish the relative importance of genes and culture.

A concern with the general approach of inferring social learning from a relationship between behavioral similarity and social relationship is that the relationship could operate in the reverse direction. Social similarity could potentially be a function of behavioral similarity if, for instance, individuals whose foraging techniques were similar tended to associate. The regression approach should indicate this, however, because if social factors are not determining behavioral differences, then genes or environment are presumably responsible. Then the standard partial regression coefficient between social similarity and behavioral similarity will be small (because genes or environment will primarily explain behavior) and social learning will not be supported, even though social and behavioral similarities are well correlated.

For more details of this regression method of inferring the extent of cultural influence on behavior, see Whitehead (In press-b).

7.6.5: Culturally Influenced Societies. Once culture takes root, societies are subject to a new range of influences (Richerson & Boyd 2004). For non-humans, the most important of these may be group conformism. Sets of animals, which could be units or communities under my definitions, develop different ways of doing things, partly through the vagaries of innovation and social learning, but also because imperatives may develop to cause individuals to behave like other members of their community and differently from members of other communities. Behavior becomes a badge of membership in a cultural community (Richerson & Boyd 1998). Operationally, how might we recognize these effects?

Conformism will tend to homogenize behavior within communities. Thus, a failure to detect difference among association rates or interaction rates within communities using the Bejder et al. (1998) test (Section 4.9) could be interpreted as the result of culturally induced behavioral conformism, but a null model of equivalence among community members (Section 6.3) is perhaps more parsimonious.

Another, and I think more powerful, sign of conformism occurs when communities actively alter their community-specific behavior to maintain distinctions. For instance, Deecke et al. (2000) studied the temporal evolution of a killer whale (*Orcinus orca*) call type in two sympatric matrilineal social units. The rate of divergence in the characteristics of the call between the two units was significantly lower than the rate of modification within either unit, showing that calls were modified in a similar fashion in the two units and suggesting that the width of the distinction between the two versions of the vocalization was important to the animals.

8 The Way Forward

Although social analysis has been something of a back-water of animal behavior science, there has been development over the last 20 years, particularly in analytical techniques and the addition of insights from genetic analysis. The rate is picking up, especially with the incorporation of network analysis into our arsenal, and I am confident that the acceleration will continue. Advances will occur in the conceptual basis of the field, the range of species examined, the data that we collect, and how we analyze it.

8.1 Conceptual Frameworks

To structure this book, I have used Hinde's (1976) seminal paper, which introduced a framework for the analysis of social structures. There are elements of social analysis, such as the role of feedback and "institutions" (e.g., conformist cultures), that he discusses but that are not yet part of the general repertoire of analyses of nonhuman social structures. As the conceptual basis of our analyses widen, however, this is beginning to change. For instance, the last 10 years has seen a great increase in interest in the role of culture in nonhuman societies (Section 7.6).

I like Hinde's framework, but it is probable that our conceptual model of social structure can be improved. Hinde's framework is based on the dyadic interaction, but interactions and resulting relationships can involve three or more animals (e.g., Kummer et al. 1974; de Waal 1998). Further in this direction, the concept of the social niche (Section 1.7) recently introduced by Flack et al. (2006) looks promising. Role theory takes an individual animal as its central focus and Hinde's framework takes the dyadic interaction, whereas the social niche combines these by focusing on the set of relationships of an individual. Because the relationships of an individual are invariably interrelated and depend on characteristics of that individual, this perspective makes a great deal of sense. How would it change our operational analyses of social structure? Not too much, I suspect, but techniques that include concepts or measures of social niche, such as network analysis (Section 5.3) and lagged association rates (Section 5.5), would come to the fore. The social niche may encourage the development of these techniques, as well as new analytical methods.

8.2 Subjects of Social Analysis

The great majority of the examples of social analysis in this book and in the wider literature are of birds or mammals, and within these groups, attention has been far from uniform. Among the mammals, attention has focused on primates, cetaceans, ungulates, and bats. There are good reasons for these foci, such as ease of study and increased interest in what we perceive as the more cognitively advanced species. However, technical advances, for instance, using devices such as PIT tags to identify individuals, and a broadening of interest is spreading the individual-based methods of social analysis to other groups. This needs to widen further. For instance, although the societies of the social insects have attracted a great deal of attention (e.g., Wilson 1971; Sumpter 2006), there has been little individual-based analysis of the type described in this book. I believe that this would illuminate much about of these fascinating animals. Conversely, whereas human societies have been the subjects of extraordinarily detailed study using a wide variety of methods (e.g., Wellman & Berkowitz 2003), some of the techniques introduced for nonhumans, such as perhaps lagged association rates (Section 5.5), might reveal new sides of our rich social structures.

8.3 Collecting Better Data

Our models of the social structure of populations (Chapter 5) as well as the power of our analyses of function and evolution (Chapter 7) are both limited by the data sets with which we have to work. Clearly, a data set with more temporal sampling periods and more individuals about whom we know more (e.g., including genetic data) and have measured a greater variety of interaction/association measures is better than one with fewer. Shear effort will increase the number of time periods, and perhaps individuals, and as field workers become more experienced, they may be able to collect more diverse data sets, for instance, by adding acoustic recordings of vocalizations to visual observations of interactions. Obtaining genetic profiles using molecular techniques (e.g., Selkoe & Toonen 2006) and using dietary profiles based on methods like fatty acid signatures (Iverson et al. 2004) are becoming easier to carry out and less expensive. Thus, we improve our picture of the nonsocial structure of the population, which may determine, or be determined by, its social structure. These incremental advances will synergistically add power to our social analyses.

In addition, we might be about to see quantum jumps in our ability to collect data on social structure in some species. In most current field and captive settings, data on only one individual or group are collected at any time (Section 3.7). With autonomous data collection, however, this limitation can be overcome. Behavior can be captured on video, by acoustic recorders, or in other media without real-time human operators, giving potentially enormous databases. However, there are two important challenges that need to be overcome for these data streams to achieve their potential. The first is to identify the participants in the behavior being recorded, and the second is to process the potentially vast amount of video, acoustic, or other data collected. Without individual identity, data on social behavior has limited value (Section 3.5). If individuals are identifiable from video images, then these data become immensely valuable, although there may be a great deal of labor-intensive work involved in obtaining the identities of the interactants as well as the interaction records from long streams of autonomously collected video. Ideally, the interaction data and identities would be abstracted from the images by computer routines, although this is probably a long way off in most cases. Acoustic data are more easily processed automatically, although even this is challenging, and identifying individuals from their acoustic signatures has only been possible in a few cases (e.g., Campbell et al. 2002; Adi et al. 2004). Sometimes identity can be determined by an additional data stream, such as the PIT tags that encode identity to special autonomous readers (Section 3.5) or acoustic tags that can encode identity and position (e.g., Voegeli et al. 2001).

More-complex tags may record interactions or signals acoustically, visually, or through other communicative modes. They thus become autonomous focal animal samplers (Section 3.7). As another example of where we might be going, McConnell et al. (2003) suggest that quite simple tags placed on animals that exchange information with each other when the animals come into proximity could quickly amass a large and detailed data set on behavioral associations. Perhaps we can go further, reliably and regularly recording physiological or neurological responses to interactions or associations with other individuals, and so rooting our model of social structure deep in the more basic biology of the animals.

8.4 Improving Analysis

With time and scientific advance, the data sets used for social analysis will grow. In fact, a few are very large at present. An example is that of the chimpanzees (*Pan troglodytes*) at Gombe, Tanzania, where a large part of the population has been followed continuously almost every day for many years (Goodall 1986). The methods described earlier in this book are certainly not sufficient to obtain anything like the most informative social model from such a data set, and some of them are just not feasible with so much data. Thus, we need special techniques to analyze such wonderfully large sources. I suspect that methods with some similarity to lagged association rates (Sections 4.6 and 5.5) may prove useful. The detailed record of interactions between two individuals might be modeled using functions that include individual factors, "personality" plus changes with age, dominance rank or reproductive state (Section 4.3), societal factors including contagion (i.e., the interactions between a dyad being triggered by the behavior of other members of their community), and the presence and attributes of individuals with significant roles such as policers or matriarchs (Section 7.1), as well as dyadic factors (Section 4.8). The individual, societal, and dyadic factors could include complex changes with time, including perhaps autoregression, so there is dependence on what has happened previously, or transient phenomena (sudden "spats" in a dyadic relationship or "periods of nervousness" in the population).

Although perhaps feasible with very large data sets, such models are very far from anything that has been achieved. In fact, only in very few cases has a statistical model been developed directly from a conceptual social model and then fitted to real data. Much more often, general mathematical models are fit to social data and then interpreted socially [e.g., in network analyses (Section 5.3) and lagged association rates (Section 5.5)]. An example of the direct fitting of a social model is instructive. The method of Durban and Parsons (In press) uses a

Bayesian framework and Markov chain Monte Carlo methods to fit quite simple models of a compartmentalized closed society formed of a number of units of unknown size, with the social assumption that individuals are more likely to associate with members of their own units (Section 5.7). The output is very useful, giving posterior distributions of numbers of units and the probabilities that pairs of individuals are members of the same unit. Currently, however, fitting this model to real data is difficult and time consuming. Improvements in hardware and software will change this and make such direct models of social structure, and more complex ones, much more useful tools.

In a number of places in this book, I suggested areas for analytical development. These include lagged interaction rates (Section 5.5), methods for disentangling environmental, genetic, and cultural influences on behavior (Section 7.6), and wavelet methods for measuring the complexity of multivariate displays (Section 6.3). The development of any of these would need some attention from a social analyst with statistical expertise or a statistician interested in the problems of social analysis.

In one area, however, we do not need to do the development ourselves. New ideas of network analysis are appearing at an extraordinary rate, driven largely by the work of a group of creative physicists (e.g., Newman 2003b; Section 5.3). Several key techniques appeared within the few months that I spent writing the network sections of this book, and undoubtedly more will arise before it is published. In this area, the social analyst needs to keep her eyes on the network literature, and then—this is the challenging part—sift the useful methods from those that add little.

And from all this? The output, from the ethological perspective, should be an increasingly clear view of the social lives of animals: how their relationships are built up, how they change, the social niches of individuals, and how all these interact to produce societies. The behavioral ecologist then has a solid base for her attacks on the "why" questions.

9 Appendices

9.1 Glossary

These definitions are for the usage in this book. Starred terms are used in various ways by different authors. For nonstarred terms, there is a generally agreed meaning for the term that is given.

Affinity The affinity of a node in a binary network is the average degree of its neighbors (Barthélemy et al. 2005), and in a weighted network, it is the weighted strength of its neighbors.∗

Aggregation Spatiotemporal clusters of individuals that are entirely the result of some nonsocial forcing factor.∗

Association Two animals are associated if their circumstances (spatial ranges, behavioral states, etc.) are those in which interactions usually take place (Whitehead & Dufault 1999).∗

Association index An estimate of the proportion of time that a pair of animals are in association (Cairns & Schwager 1987).

Assortativity The extent to which nodes in a network are connected to nodes that are similar to themselves (Newman 2003b).

Asymmetric relationship A dyadic relationship in which the members interact with one another at significantly different rates.

Behavioral ecology The study of functional questions about animal behavior (Krebs & Davies 1991).

Bond Two animals are bonded if they have a consistently strong relationship in two or more independent behavioral modes (R. Wrangham, personal communication).∗

Clustering coefficient In a network, the extent to which the nodes connected to a focal node are themselves connected (Newman 2003b).

Community A set of individuals that is largely behaviorally self-contained and within which most individuals interact with most others (Goodall 1968).∗

Culture Information or behavior shared by a population or subpopulation that is acquired from conspecifics through some form of social learning (Rendell & Whitehead 2001).∗

Degree In a binary network analysis, the number of other nodes connected to a node.∗

Dendrogram Tree diagram, in which individuals are represented by nodes and the branching pattern indicates degrees of association, the results of a hierarchical cluster analysis.

Dependency Describes a situation in which an animal probably would not survive without the behavior of another.∗

Dominance A consistent outcome in favor of one member of a dyad during repeated, agonistic interactions between two individuals, and a default yielding of the other individual rather than escalation (Drews 1993).

Dominance hierarchy An ordering of individuals such that more highly ranked individuals generally win agonistic encounters over, or receive submissive behavior from, those ranked lower.

Dominance index A measure of an individual's ability to dominate others in its community.

Dominance rank The ranking of an in individual within its community in its ability to consistently win repeated agonistic encounters with other members of the community.

Dyadic mode A method of storing data in which each row corresponds to an observation of a dyad, so that there are two identity fields representing the two individuals in the dyad (as opposed to group or linear mode).

Edge Relationship between two nodes in a network.∗

Eigenvector centrality In a network analysis, eigenvector centrality is a measure of how well connected a node is

(Newman 2004). Mathematically, it is the first eigenvector of the matrix of edges or weights.

Equivalence Things, including social partners, that become mutually interchangeable through common spatiotemporal or functional interactions (Schusterman et al. 2000).*

Ethology The study of animal behavior based on description (Hinde 1982).

Fission–fusion "A society consisting of casual groups of variable size and composition, which form, break-up and reform at frequent intervals" (Conradt & Roper 2005).

Follow A research strategy in which the researcher's attention stays with an individual or group (as opposed to a survey).*

Gregariousness Mean number of associates possessed by an individual (Pepper et al. 1999).

Group Sets of animals that actively achieve or maintain spatiotemporal proximity over any time scale and within which most interactions occur.*

Group mode A method of storing data in which observations of more than one individual are represented on each row in one field (as opposed to dyadic or linear mode).

Inconsistency In a dominance hierarchy, the situation in which a lower-ranking individual dominates a more highly ranked one.*

Interaction An action of one animal directed toward another or affecting the behavior of another.*

Kinship Genetic relatedness through common ancestry.

Lagged association rate The probability that a dyad is associated at some time after a recorded association (Whitehead 1995).

Likelihood The probability of obtaining a data set, given a particular model and set of parameters.

Linear mode A method of storing data in which observations of one individual are represented on each row (as opposed to dyadic or group mode).

Linearity A measure of the consistency with which individuals higher in a dominance hierarchy dominate those ranked lower, usually measured by Landau's (1951) h.*

Mantel test Permutation test of the significance of the relationship between the corresponding, nondiagonal, elements of two similarity or dissimilarity matrices indexed by the same individuals, with the null hypothesis being that there is no relationship (Mantel 1967).

Matrix correlation The correlation coefficient between the corresponding, nondiagonal, elements of two similarity or dissimilarity matrices indexed by the same individuals, ranging from −1, indicating a perfect negative linear relationship, to 0, no relationship, to 1, perfect positive linear relationship.

Modularity For some arrangement of individuals into clusters, the difference between the proportion of the total association within clusters and the expected proportion when individuals associate randomly (Newman 2004).∗

Network Pattern of connectedness among members of a population (Newman 2003b).

Node Vertex of a network (usually an individual animal in social analysis).∗

Null association rate The expected probability that a dyad is associated at some time after a recorded association if association had no time dependence (Whitehead 1995).

Ordination Visual display in which points represent individuals and their proximity to one another indicates their degree of association.

Population A set of individuals, usually of the same species, such that the great majority of interactions involving members take place with other members of the population.∗

Reach A measure of indirect connectedness in a network such that nodes with high reach are connected indirectly to other nodes of high degree or strength (Flack et al. 2006).∗

Reciprocity Situation in which an increase in the rate at which behavior of a particular type by one individual toward another increases the rate at which the same type of behavior is reciprocated (Hemelrijk 1990b). In "reputation reciprocity," individuals have increased interaction rates with those whose overall interaction rate is high (Mohtashemi & Mui 2003). In "generalized reciprocity," individuals interact at rates correlated with the rates at which they are interacted with (Pfeiffer et al. 2005).

Relationship A synthesis of the content, quality, and patterning of the interactions between two individuals, where patterning is with respect to both each others' behavior and time (Hinde 1976).∗

Relationship measures Quantitative descriptors of the content, quality, or temporal patterning of dyadic relationships (Whitehead & Dufault 1999).

Social differentiation The degree to which the dyads within a population differ in their probability of association, measured using an estimate of the coefficient of variation of the true association index.

Social niche A vector of behavioral connections in the set of overlapping social networks in which an individual participates (Flack et al. 2006).

Social organization Synonymous with social structure.*

Social structure A synthesis of the nature, quality, and patterning of the relationships among the members of a population (Hinde 1976). In this book, social structure is treated as synonymous with social organization, social system, and society.*

Social system Synonymous with social structure.*

Social unit (unit). Set of animals that are permanently or nearly permanently in association (Whitehead 1995).*

Society Synonymous with social structure.*

Sociogram Diagrammatic representation of social structure in which individuals are represented by nodes, and edges between nodes indicate the strength of the dyadic relationship.*

Steepness The certainty with which a dominant wins an interaction over a subordinate in a dominance hierarchy (de Vries et al. 2006).*

Strength In a weighted network analysis, the sum of the weights of the edges connected to a node (Barthélemy et al. 2005).*

Survey A research strategy in which an individual or group is first encountered and then observed, and then the researcher moves on to another individual or group (as opposed to a follow).*

Symmetric A relation is symmetric if, when A relates to B, then it necessarily follows that B relates to A.*

Tier Level of a hierarchical social organization in which elements of tier i social entities are tier $i - 1$ social entities (Wittemyer et al. 2005).*

Transitive A relation is transitive if, when A relates to B, and B relates to C, then it necessarily follows that A relates to C.

In social analysis, transitive associations define groups and transitive dominance relationships define perfect dominance hierarchies.*

Typical group size The mean group size that an individual or set of individuals experiences (Jarman 1974).

9.2 A Key Journal and Some Useful Books

The journal *Animal Behaviour*, jointly published by the Animal Behavior Society (United States) and the Association for the Study of Animal Behaviour (United Kingdom), contains a large proportion of the key papers on the methods of analysis of nonhuman societies, as well as many good examples.

Here are some of the books that I find most useful when analyzing animal societies:

Bradbury, J. W., and S. L. Vehrencamp. 1998. *Principles of animal communication*. Sunderland, MA: Sinauer Associates. A wide-ranging and clear consideration of animal communication.

Burnham, K. P., and D. R. Anderson. 2002. *Model selection and multimodel inference: a practical information-theoretic approach*, 2nd ed. New York: Springer-Verlag. Explains in detail the justification and practice of model selection using AIC and related methods.

Legendre, P., and L. Legendre. 1998. *Numerical ecology*, 2nd ed. Amsterdam: Elsevier. Intended for ecologists, this book does a good job of describing techniques in ordination, classification, and related data analyses.

Lehner, P. N. 1998. *Handbook of ethological methods*, 2nd ed. Cambridge: Cambridge University Press. A great deal of useful information on how to study animal behavior.

Manly, B. F. J. 1994. *Multivariate statistical methods*, 2nd ed. New York: Chapman and Hall. A short and particularly clear introduction to multivariate statistical methods.

Manly, B. F. J. 1997. *Randomization, bootstrap and Monte Carlo methods in biology*, 2nd ed. London: Chapman and Hall. Describes the rationale for permutation methods and methodologies and some potential problems.

Martin, P., and P. Bateson. 2007. *Measuring behaviour: an introductory guide*, 3rd ed. Cambridge: Cambridge

University Press. A widely cited and very clear summary
of how to study animal behavior.

Sokal, R. R., and F. J. Rohlf. 1994. *Biometry*, 3rd ed. New York:
W. H. Freeman. This is the classic compendium of the principal statistical methods used in the biological sciences.

9.3 Computer Programs

Useful computer programs are introduced in Section 2.9 and summarized in Table 2.6. Here, I first outline which programs can perform
the general analytical tasks that are often needed in social analysis, and
then I list those that carry out some of the more specialized analyses
described in other sections of this book. Where appropriate, I indicate,
following the name of the program, the section of the program manual
where the method is described. As new versions of the computer packages are introduced, however, these references to manual sections will
become inaccurate.

9.3.1: General Tasks

Matrix manipulation. Matrices can be manipulated by Pop-
Tools, MatMan (3), and SOCPROG (7.2.2), as well as by the
general statistical analysis packages (Table 2.6). MATLAB
is extraordinarily powerful and versatile at matrix manipulation.

Permutation tests. Permutation tests can be programmed in the
more sophisticated of the general statistical analysis packages (especially S-PLUS and R) and in MATLAB, and are
relatively easily accomplished using PopTools. The Bejder
et al. (1998; Section 4.9) permutation tests and their variants
are available in SOCPROG (5.16).

Mantel tests, partial Mantel tests, and so on. These important
matrix permutation tests are carried out by PopTools,
MatMan (4.4, 4.5), R, and SOCPROG (7.4.4).

9.3.2: Specialized Tasks (Indexed to Sections of Chapters 4, 5, and 7)

Nonsocial measures of relationship (Section 4.2). Transforming
a list of attributes of individuals (e.g., sex, age) into a
matrix of their similarities or dissimilarities can be performed using the general matrix manipulation techniques

noted in the foregoing list or by SOCPROG (7.2.1). Genetically derived kinship measures between individuals are produced by programs like KINSHIP (Queller
& Goodnight 1989; gsoft.smu.edu/GSoft.html). Measures of range overlap are produced by geographical
information systems (GIS) software such as ArcView
(www.esri.com/software/arcgis/arcview/index.html).

Gregariousness (Section 4.3). Once one has a matrix of association indices (see later discussion), then the gregariousness
of the individuals is calculated simply by summing across
rows or down columns. Note that the diagonals of the association index matrix should be 0. If they are 1, then 1 should
be subtracted from each sum. In this way, gregariousness can
be calculated easily by spreadsheet programs such as Excel
as well as any of the matrix manipulation routines noted
earlier. It is also calculated by SOCPROG (5.8). SOCPROG
(5.16.6) can also test for differences in gregariousness
between individuals using permutation tests.

Rates of interaction (Section 4.4). The first step in calculating
dyadic rates of interaction is often to produce a matrix of
the total numbers of interactions between each dyad. Obtaining these sums from records of interactions can be a
little complex in many programs (e.g., Excel). However,
the use of dyadic mode for coding the data may make this
simpler because standard "pivot table" commands can sum
interaction numbers for each dyad. SOCPROG (5.1; use
"Sum of associations" option when choosing an association
index) will calculate interaction rates if the data are coded
in linear or group mode as suggested in Section 3.8. A second step, often necessary, is to obtain a measure of effort for
each dyad (Section 3.8); this is done similarly to calculating
the total numbers of interactions. The matrix of numbers of
interactions is then divided by the effort matrix to give interaction rates, a step that is easily done by any of the matrix
manipulation routines noted here. Alternatively, all this can
be done at once by writing a "Custom association index" in
SOCPROG (5.1.1).

Association indices (Section 4.5). Association indices can be
calculated from coded association data in a similar way to
that just outlined for rates of interaction, although this is
cumbersome. If we already have matrices that give the totals

of numbers of associations and numbers of identifications
of the individuals, then association indices can be calculated
easily using any of the general matrix manipulation methods
noted. Alternatively, and easier, is to calculate association
indices directly from coded data using SOCPROG (5.1).
SOCPROG (5.4) will also estimate the precision of association indices, using analytical [Equation (6)] or bootstrap
methods.

Dyadic lagged association rates (Section 4.6). Lagged association rates and related measures (null rates, etc.) are calculated by SOCPROG (6). To obtain these for a particular
dyad, use SOCPROG's (3.2) ability to restrict the data to the
two individuals of interest.

Multivariate descriptions (Sections 4.7 and 5.6). If we have several matrices of interaction rates or association indices, these
can be arranged into a single multivariate data set indexed
by dyads using the matrix manipulation routines noted.
Alternatively, SOCPROG (7) will do this, and it will also
edit the matrices so that only individuals present in all of
them are included. In this way, if one type of association was
calculated for one set of individuals and another type for a
different but overlapping set, the multivariate data set only
includes individuals in both original matrices. The multivariate data set can be output from SOCPROG (7.4.3) and
then used in any standard statistical software (Table 2.6).
SOCPROG (7.4) can do a few analyses on such data sets,
including principal components analysis.

General and generalized linear models of interaction rates (Section 4.8). General linear models can be fitted using any of
the standard statistical packages (Table 2.6). For generalized
linear models, I recommend the more sophisticated packages, especially S-PLUS and R.

Measures of asymmetry (Section 4.8). Beilharz and Cox's
(1967) measure of asymmetry or de Vries et al.'s (2006)
dyadic dominance index can be calculated using any of the
matrix manipulation routines noted.

Analyses of reciprocity (Section 4.8). MatMan (4.4, 4.5, 4.6)
and SOCPROG (5.17) perform analyses of reciprocity.

Tests for preferred/avoided companions (Section 4.9). The
Bejder et al. (1998) test and its variants are performed by
SOCPROG (5.16).

Interclass association indices and interaction rates (Section
    4.11). Mean interaction rates and association indices be-
    tween classes (such as males and females), as well as stan-
    dard errors, are calculated by SOCPROG (5.8). Also output
    are the results of Mantel tests for differences in association
    indices or interaction rates within versus between classes.
    SOCPROG will perform many of its analyses either within
    or between particular classes of individual.

Social differentiation (Section 5.1). Social differentiation and a
    bootstrap standard error are estimated by SOCPROG (4),
    using both the maximum likelihood and Poisson approxima-
    tion methods.

Histograms of association indices or interaction rates (Section
    5.2). These can be produced by any statistical software
    package, although it is necessary to remove diagonal entries
    (associations/interactions of individuals with themselves)
    and, usually, to rearrange the square association or interac-
    tion matrix. SOCPROG (5.9) does these tasks automatically
    when producing histograms.

Sociograms (Section 5.2). The best sociograms that I have
    found are those produced by NetDraw. There are many
    options, and the plot can be easily adjusted by moving
    nodes with the mouse. SOCPROG (5.12) makes simple
    sociograms.

Principal coordinates analyses and multidimensional scaling
    (Section 5.2). Most standard statistical packages (Table
    2.6) and some network software (e.g., NetDraw) include
    these ordinations, although sometimes (e.g., in SYSTAT)
    principal coordinates analyses have to be procured using
    options for the principal components analysis commands.
    SOCPROG gives both principal coordinates (5.13) and
    multidimensional scaling (5.14) ordinations.

Cluster analysis (Section 5.2). Cluster analysis is a standard part
    of all statistical packages (e.g., Table 2.6). However, not all
    give the very useful cophenetic correlation coefficient (CCC).
    SOCPROG (5.15) provides several types of hierarchical
    cluster analysis, outputs the CCC, can save the clusters pro-
    duced, and has other potentially useful options, such as the
    output of modularity and knot diagrams.

Network analysis (Section 5.3). A list of some popular network
    software is given in Table 2.6. For a more comprehensive

list, see www.insna.org/INSNA/soft_inf.html. UCINET is a favorite with many, and NetDraw does an excellent job of allowing the visualization of networks. SOCPROG (5.7) will output matrices of association indices or interaction rates in .vna format that UCINET and NetDraw can use. David Lusseau and I have added a number of weighted network statistics, plus bootstrap standard errors and permutation tests, to the output options of SOCPROG (5.10).

Dominance hierarchies (Section 5.4). MatMan (4.1, 4.2, 4.3) carries out analyses of dominance hierarchies, giving statistics such as Landau's h and de Vries' h', and ranking the individuals using the I&SI method. PeckOrder uses raw data (dyadic win–lose encounter results) or summary data to analyze dominance hierarchies, and outputs include Landau's h and won–lost records of dyads.

Lagged association rates (Section 5.5). In its temporal analysis module, SOCPROG (6) calculates lagged, null, and intermediate association rates, with a number of options, including interclass rates, different smoothing moving averages, model fitting, and jackknife standard errors. Lagged identification rates are calculated in a separate module of SOCPROG (10.1.2).

Multivariate methods (Section 5.6). See *Multivariate descriptions* (Section 4.7).

Delineating communities, units, and so on (Section 5.7). The SOCPROG (5.15) hierarchical cluster analysis function has options for the delineation of communities, units, tiers, and so on using either maximum modularity or knots. Also available in SOCPROG (5.11) is Newman's (2006) eigenvector method of dividing communities using either modularity-G or modularity-P. A variety of modular division techniques are available in the network analysis packages such as UCINET and NetDraw. There are no software packages available for the delineation of communities or other social tiers using the multivariate, temporal, or Bayesian methods described in the text (although the Bayesian techniques use WINBUGS software; www.mrc-bsu.cam.ac.uk/bugs/).

Method of independent contrasts (Section 7.4). The program CAIC is usually used for the analysis of data using the method of independent contrasts (http://www.bio.ic.ac.uk/evolve/software/caic/).

9.4 Estimating Social Differentiation

Social differentiation is the coefficient of variation of the true association indices (α'–the prime is to distinguish the true association index from its estimated value), $S = \text{CV}.(\alpha'_{IJ})$ [$I \neq J$]. x_{IJ}, the number of observations of individuals I and J associated, can be considered binomially distributed with coefficient α'_{IJ}, their true association index, and number of samples d_{IJ}, the denominator of the estimated association index (as given in Table 4.5). I assume that the α_{IJ} are distributed according to the beta distribution, which gives values between 0 and 1, with mean μ and coefficient of variation S. Then the parameters of the beta distribution are. $\beta_1 = \mu \cdot [(1-\mu)/(\mu \cdot S^2) - 1]$ and. $\beta_2 = (1-\mu) \cdot [(1-\mu)/(\mu \cdot S^2) - 1]$, and the likelihood of the data, the $\{x_{IJ}\}$, given μ and S is proportional to (Whitehead In press-a)

$$L = \prod_{IJ} \int_0^1 {\alpha'_{IJ}}^{x_{IJ}} \cdot (1 - \alpha'_{IJ})^{(d_{IJ} - x_{IJ})} \cdot \text{B}(\alpha'_{IJ}, \beta_1, \beta_2) \cdot \text{d}(\alpha'_{IJ}) \quad (23)$$

where $\text{B}(\alpha'_{IJ}, \beta_1, \beta_2)$ is the probability density function of the beta distribution with parameters β_1 and β_2 at α'_{IJ}. We then choose $\{\alpha'_{IJ}\}$, μ and S to maximize L. The maximization and the integration in Equation (23) are done numerically by SOCPROG.

A simpler method of estimating social differentiation is obtained if we assume that the probability that an individual is identified in a sampling period is the same for all individuals in all sampling periods and the number of observed associations of any pair is Poisson distributed. With these assumptions, social differentiation can be estimated by

$$S = \frac{\sqrt{\text{Var}(x_{IJ}) - \text{Mean}(x_{IJ})}}{\text{Mean}(x_{IJ})} \quad (24)$$

Simulations indicate that this Poisson estimate of S [Equation (24)] is a rough approximation of real social differentiation but is often considerably more biased and less precise than the maximum likelihood estimator [Equation (23)].

9.5 Assessing Unit Size, Group Size, or Community Size

In ideal circumstances, the sizes of social entities can be enumerated directly from the number of identified individuals in each. Thus, if their memberships can be determined (Section 5.7), then size can be enumerated. In other situations, however, we need to use less direct methods.

For groups that are reasonably well delineated spatiotemporally, a number of the standard population assessment techniques may be useful. For instance, line or strip transect methodology can be used to assess the numbers of animals in large and/or dispersed groups, with the censuses being carried out using any of a variety of platforms and detection methods. These include observers walking transect lines with binoculars, use of aerial photographs, or use of acoustic surveys with hydrophones towed by boats. Numbers of animals or cues (such as calls, scent markings, or highly visible activities) counted are extrapolated to group sizes by accounting for individuals missed through not surveying the entire range of the group because some animals were not available to be counted (e.g., hiding or underwater) and, in the case of cue counts, the rates at which animals produce the cues. There is a well-developed methodology for such censuses that is described by Buckland et al. (1993).

Mark-recapture is another standard population assessment technique that can be used to estimate group size. The concept is straightforward. If the animals are observed in two sampling periods and those identified are recorded, the proportion of previously identified individuals observed during the second period should be roughly the ratio of the number identified during the first period to the group size. There are many elaborations of this, which consider more than two sampling periods, as well as birth, death, emigration, and immigration (Seber 1982, 1992). For estimating the size of a closed group, however, the simplest two-sample "Petersen" scenario is often sufficient. There are three important assumptions: that the group is closed; that all animals are equally identifiable; and that identification in the first sample does not make the animal more or less likely to be identified in the second. If these conditions hold, then the group size N_g can be estimated from the following formula, which is a slight elaboration of the proportional argument given earlier to correct for bias (Seber 1982, p. 60):

$$N_g = \frac{(n_1 + 1)(n_2 + 1)}{(m + 1)} - 1$$

where n_1 is the number of animals identified in the first sample, n_2 is the number of animals identified in the second sample, and m is the number of individuals identified in both samples. Confidence intervals of such estimates can be obtained using binomial, Poisson, or normal-approximation methods (e.g., Krebs 1989, pp. 17–22) or bootstrap methods. An approximate standard error is (Seber 1982, 60)

$$\text{SE}(N_g) = \sqrt{\frac{(n_1 + 1)(n_2 + 1)(n_1 - m)(n_2 - m)}{(m + 1)^2(m + 2)}}$$

Assessments using this method might need to be corrected if some animals are not identifiable (Section 3.5).

Whereas, as I have defined them, groups are spatiotemporally discrete, units and communities need not be. If a unit or community is known to be well delineated in space, then the same line transect and mark-recapture techniques are applicable. If it is not, for instance, if a unit frequently mixes with other units or a community shares its range with another, the best way of estimating its size is to use the methods described in Section 5.7 to delineate the unit or community membership and then count the members. If some individuals are not identifiable and so are not included in the enumeration, then a correction may be needed (Section 3.5).

If we are more interested in the overall distribution of group, community, or unit sizes in a population, then shortcut methods are available. Fitting models to lagged association rates can give estimates of typical group size and/or typical unit size (as described in Section 5.5), with temporal jackknives indicating standard errors.

The size of the social unit containing an individual A can be estimated if the population is assumed to be large and we have data from three widely spaced sampling periods, so that common companions in two of the periods can be assumed to be fellow unit members (Christal et al. 1998):

$$N_u(A) = \frac{(n_{12}(A) + 1)(n_{23}(A) + 1)(n_{13}(A) + 1)}{(n_{123}(A) + 1)^2} \tag{25}$$

where $n_{12}(A)$ is the number of common associates of A in sampling periods 1 and 2, and $n_{123}(A)$ is the number of common associates of A in all three sampling periods.

The estimates from lagged association rates and the mean unit sizes estimated using Equation (25) all give mean typical group and unit sizes, that is, those experienced by a member of the population (Jarman 1974). These are at least as great as those experienced by outside observers because there are more individuals in larger groups (Section 3.4).

References

Adi, K., T. S. Osiejuk, and M. T. Johnson. 2004. Automatic song-type classification and individual identification of the ortolan bunting (*Emberiza hortulana* L) bird vocalizations. *J. Acoust. Soc. Am.* 116:2639.

Akaike, H. 1973. Information theory as an extension of the maximum likelihood principle. In *Second international symposium on information theory*, ed. B. N. Petrov and F. Csaki. Budapest: Akademiai Kiado, pp. 267–281.

Albers, P. C. H., and H. de Vries. 2001. Elo-rating as a tool in the sequential estimation of dominance strengths. *Anim. Behav.* 61:489–495.

Altmann, J. 1974. Observational study of behavior: sampling methods. *Behaviour* 49:227–267.

Altmann, S. A., and J. Altmann. 1977. On the analysis of rates of behaviour. *Anim. Behav.* 25:364–372.

Appleby, M. C. 1983. The probability of linearity in hierarchies. *Anim. Behav.* 31:600–608.

Araabi, B. N., N. Kehtarnavaz, T. McKinney, G. R. Hillman, and B. Würsig. 2000. A string matching computer-assisted system for dolphin photo-identification. *Ann. Biomed. Eng.* 28:1269–1279.

Archie, E. A., C. J. Moss, and S. C. Alberts. 2006. The ties that bind: genetic relatedness predicts the fission and fusion of social groups in wild African elephants. *Proc. R. Soc. Lond. B* 273:513–522.

Axelrod, R., and W. D. Hamilton. 1981. The evolution of cooperation. *Science* 211:1390–1396.

Baird, R. W., and H. Whitehead. 2000. Social organization of mammal-eating killer whales: group stability and dispersal patterns. *Can. J. Zool.* 78:2096–2105.

Baker, M. C., and S. F. Fox. 1978. Dominance, survival, and enzyme polymorphism in dark-eyed juncos, *Junco hyemalis*. *Evolution* 32:697–711.

Barabási, A.-L., and R. Albert. 1999. Emergence of scaling in random networks. *Science* 286:509–512.

Barkow, J. H. 2001. Culture and hyperculture: why can't a cetacean be more like a (hu)man? *Behav. Brain Sci.* 24:324–325.

Barrat, A., M. Barthélemy, R. Pastor-Satorras, and A. Vespignani. 2004. The architecture of complex weighted networks. *Proc. Natl. Acad. Sci. USA* 101:3747–3752.

Barrett, L., P. Henzi, and R. Dunbar. 2003. Primate cognition: from 'what now?' to 'what if?'. *Trends Cog. Sci.* 7:494–497.

Barthélemy, M., A. Barrat, R. Pastor-Satorras, and A. Vespignani. 2005. Characterization and modeling of weighted networks. *Physica A* 346:34–43.

Bayly, K. L., C. S. Evans, and A. Taylor. 2006. Measuring social structure: a comparison of eight dominance indices. *Behav. Proc.* 73:1–12.

Beilharz, R. G., and D. F. Cox. 1967. Social dominance in swine. *Anim. Behav.* 15:117–122.

Bejder, L., D. Fletcher, and S. Bräger. 1998. A method for testing association patterns of social animals. *Anim. Behav.* 56:719–725.

Bigg, M. A., P. F. Olesiuk, G. M. Ellis, J. K. B. Ford, and K. C. Balcomb. 1990. Social organization and genealogy of resident killer whales (*Orcinus orca*) in the coastal waters of British Columbia and Washington State. *Rep. Int. Whal. Commn. (Spec. Iss.)* 12:383–405.

Blumstein, D. T., and J. C. Daniel. 2007. *Quantifying behavior the J Watcher way*. Sunderland, MA: Sinauer Associates.

Blumstein, D. T., and K. B. Armitage. 1998. Life history consequences of social complexity: a comparative study of ground-dwelling sciurids. *Behav. Ecol.* 9:8–19.

Boccaletti, S., V. Latora, Y. Moreno, M. Chavez, and D.-U. Hwang. 2006. Complex networks: structure and dynamics. *Phys. Rep.* 424:175–308.

Boesch, C. 2001. Sacrileges are welcome in science! Opening a discussion about culture in animals. *Behav. Brain Sci.* 24:327–328.

Boesch, C., and H. Boesch-Achermann. 2000. *The chimpanzees of the Taï Forest. Behavioural ecology and evolution*. Oxford: Oxford University Press.

Borgatti, S. P., M. G. Everett, and L. C. Freeman. 1999. *UCINET 6.0 Version 1.00*. Natick, MA: Analytic Technologies.

Boyd, R., and J. B. Silk. 1983. A method for assigning cardinal dominance ranks. *Anim. Behav.* 31:45–58.

Bradbury, J. W., and S. L. Vehrencamp. 1998. *Principles of animal communication*. Sunderland, MA: Sinauer Associates.

Bridge, P. D. 1993. Classification. In *Biological data analysis*, ed. J. C. Fry. Oxford: Oxford. University Press, pp. 219–242.

Brown, J. L. 1975. *The evolution of behavior*. New York: Norton.

Brownell, R. L., and K. Ralls. 1986. Potential for sperm competition in baleen whales. *Rep. Int. Whal. Commn. (Spec. Iss.)* 8:97–112.

Buckland, S. T., D. R. Anderson, K. P. Burnham, and J. L. Laake. 1993. *Distance sampling: estimating abundance of biological populations.* New York: Chapman and Hall.

Burnham, K. P., and D. R. Anderson. 2002. *Model selection and multimodel inference: a practical information-theoretic approach*, 2nd ed. New York: Springer-Verlag.

Byrne, R., and A. Whiten. 1988. *Machiavellian intelligence.* Oxford: Clarendon.

Cairns, S. J., and S. J. Schwager. 1987. A comparison of association indices. *Anim. Behav.* 35:1454–1469.

Campbell, G. S., R. C. Gisiner, D. A. Helweg, and L. L. Milette. 2002. Acoustic identification of female Steller sea lions (*Eumetopias jubatus*). *J. Acoust. Soc. Am.* 111:2920–2928.

Carlin, B. P., and T. A. Louise. 2000. *Bayes and empirical Bayes methods for data analysis*, 2nd ed. New York: Chapman and Hall.

Chambers, L. K., G. R. Singleton, and C. J. Krebs. 2000. Movements and social organization of wild house mice (*Mus domesticus*) in the wheatlands of northwestern Victoria, Australia. *J. Mammal.* 81:59–69.

Chase, I. D., C. Tovey, D. Spangler-Martin, and M. Manfredonia. 2002. Individual differences versus social dynamics in the formation of animal dominance hierarchies. *Proc. Natl. Acad. Sci. USA* 99:5744–5749.

Christal, J., and H. Whitehead. 2001. Social affiliations within sperm whale (*Physeter macrocephalus*) groups. *Ethology* 107:323–340.

Christal, J., H. Whitehead, and E. Lettevall. 1998. Sperm whale social units: variation and change. *Can. J. Zool.* 76:1431–1440.

Clutton-Brock, T. H. 1989. Mammalian mating systems. *Proc. R. Soc. Lond. B* 236:339–372.

Clutton-Brock, T. H., F. E. Guinness, and S. D. Albon. 1982. *Red deer. Behavior and ecology of two sexes.* Chicago: University of Chicago Press.

Clutton-Brock, T. H., S. D. Albon, and F. E. Guinness. 1979. The logical stag: adaptive aspects of fighting in red deer (*Cervus elaphus* L). *Anim. Behav.* 27:211–225.

Cohen, J. E. 1971. *Casual groups of monkeys and men.* Cambridge, MA: Harvard University Press.

Cole, L. C. 1949. The measurement of interspecific association. *Ecology* 30:411–424.

Connor, R. C. 2000. Group living in whales and dolphins. In *Cetacean societies*, ed. J. Mann, R. C. Connor, P. L. Tyack, and H. Whitehead. Chicago: University of Chicago Press, pp. 199–218.

Connor, R. C., R. Smolker, and L. Bejder. 2006. Synchrony, social behaviour and alliance affiliation in Indian Ocean bottlenose dolphins, *Tursiops truncatus. Anim. Behav.* 72:1371–1378.

Connor, R. C., M. R. Heithaus, and L. M. Barre. 2001. Complex social structure, alliance stability and mating access in a bottlenose dolphin 'super-alliance'. *Proc. R. Soc. Lond. B.* 268:263–267.

Connor, R. C., R. S. Wells, J. Mann, and A. J. Read. 2000. The bottlenose dolphin. Social relationships in a fission–fusion society. In *Cetacean societies*, ed. J. Mann, R. C. Connor, P. L. Tyack, and H. Whitehead. Chicago: University of Chicago Press, pp. 91–126.

Connor, R. C., J. Mann, P. L. Tyack, and H. Whitehead. 1998. Social evolution in toothed whales. *Trends Ecol. Evol.* 13:228–232.

Connor, R. C., R. A. Smolker, and A. F. Richards. 1992. Two levels of alliance formation among male bottlenose dolphins (*Tursiops* sp.). *Proc. Natl. Acad. Sci. USA* 89:987–990.

Conradt, L., and T. J. Roper. 2005. Consensus decision making in animals. *Trends Ecol. Evol.* 20:449–456.

Corradino, C. 1990. Proximity structure in a captive colony of Japanese monkeys (*Macaca fuscata fuscata*): an application of multidimensional scaling. *Primates* 31:351–362.

Costa, J. T., and T. D. Fitzgerald. 1996. Developments in social terminology: semantic battles in a conceptual war. *Trends Ecol. Evol.* 11:285–289.

Croft, D. P., R. James, and J. Krause. In press. *Exploring animal social networks.* Princeton: Princeton University Press.

Croft, D. P., R. James, A. J. W. Ward, M. S. Botham, D. Mawdsley, and J. Krause. 2005. Assortative interactions and social networks in fish. *Oecologia* 143:211–219.

Croft, D. P., J. Krause, and R. James. 2004. Social networks in the guppy (*Poecilia reticulata*). *Proc. R. Soc. Lond. B* 271:S516–S519.

Crook, J. H. 1970. Introduction—-social behaviour and ethology. In *Social behaviour in birds and mammals. Essays on the social ethology of animals and man*, ed. J. H. Crook. London: Academic Press. pp. xxi–l.

Crook, J. H., and P. A. Butterfield. 1970. Gender role in the social system of Quelea. In *Social behaviour in birds and mammals. Essays on the social ethology of animals and man*, ed. J. H. Crook. London:de Academic Press, pp. 211–248.

Crook, J. H., and J. S. Gartlan. 1966. Evolution of primate societies. *Nature* 210:1200–1203.

Crow, E. L. 1990. Ranking paired contestants. *Commun. Statist. Simulation* 19:749–769.

da Silva, J., and J. M. Terhune. 1988. Harbour seal grouping as an anti-predator strategy. *Anim. Behav.* 36:1309–1316.

David, H. A. 1987. Ranking from unbalanced paired-comparison data. *Biometrika* 74:432–436.

Dawkins, R. 1976. *The selfish gene.* Oxford: Oxford University Press.

Deecke, V. B., J. K. B. Ford, and P. Spong. 2000. Dialect change in resident killer whales: implications for vocal learning and cultural transmission. *Anim. Behav.* 40:629–638.

de Vries, H. 1998. Finding a dominance order most consistent with a linear hierarchy: a new procedure and review. *Anim. Behav.* 55:827–843.

———1995. An improved test of linearity in dominance hierarchies containing unknown or tied relationships. *Anim. Behav.* 50:1375–1389.

de Vries, H., J. M. G. Stevens, and H. Vervaecke. 2006. Measuring and testing the steepness of dominance hierarchies. *Anim. Behav.* 71:585–592.

de Vries, H., and M. C. Appleby. 2000. Finding an appropriate order for a hierarchy: a comparison of the I&SI and BBS methods. *Anim. Behav.* 59:239–245.

de Vries, H., W. J. Netto, and P. L. H. Hanegraaf. 1993. MatMan: a program for the analysis of sociometric matrices and behavioural transition matrices. *Behaviour* 125:157–175.

de Waal, F. 1998. *Chimpanzee politics. Power and sex among apes*, 2nd ed. Baltimore: Johns Hopkins University Press.

Dietz, E. J. 1983. Permutation tests for association between two distance matrices. *Syst. Zool.* 32:21–26.

Digby, P. G. N., and R. A. Kempton. 1987. *Multivariate analysis of ecological communities*. London and New York: Chapman and Hall.

Dobson, A. J. 2001. *An introduction to generalized linear models*, 2nd ed. London: Chapman and Hall/CRC.

Drews, C. 1993. The concept and definition of dominance in animal behaviour. *Behaviour* 125:283–311.

Dunbar, R. I. M. 1998. The social brain hypothesis. *Evol. Anthropol.* 6:178–190.

Dunstan, F. D. J. 1993. Time series analysis. In *Biological data analysis*, ed. J. C. Fry. Oxford: Oxford. University Press. pp. 243–310.

Durban, J. W., and K. M. Parsons. In press. Quantifying clusters in social populations. *Behav. Ecol.*

Durrell, J. A., I. A. Sneddon, N. E. O'Connell, and H. Whitehead. 2004. Do pigs form preferential associations? *Appl. Anim. Behav. Sci.* 89:41–52.

Ebensberger, L. A. 1998. Strategies and counterstrategies to infanticide in mammals. *Biol. Rev.* 73:321–346.

Edwards, A. W. F. 1992. *Likelihood*, 2nd ed. Baltimore: Johns Hopkins University Press.

Efron, B., and G. Gong. 1983. A leisurely look at the bootstrap, the jackknife, and cross-validation. *Am. Stat.* 37:36–48.

Efron, B., and C. Stein. 1981. The jackknife estimate of variance. *Ann. Stat.* 9:586–596.

Emlen, S. T., and L. W. Oring. 1977. Ecology, sexual selection, and the evolution of mating systems. *Science* 197:215–223.

Espinas, A. 1878. *Des sociétés animales: étude de psychologie comparée*. Paris: Libraire Germer Ballière.

Faulkes, C. G., N. C. Bennett, M. W. Bruford, H. P. O'Brien, G. H. Aguilar, and J. U. M. Jarvis. 1997. Ecological constraints drive social evolution in the African mole-rats. *Proc. R. Soc. Lond. B* 264:1619–1627.

Faust, K., and J. Skvoretz. 2002. Comparing networks across space, time and species. *Sociol. Methodol.* 32:267–299.

Felsenstein, J. 1985. Phylogenies and the comparative method. *Am. Nat.* 125:1–15.

Ficken, M. S., S. R. Witkin, and C. M. Weise. 1981. Associations among members of a black-capped chickadee flock. *Behav. Ecol. Sociobiol.* 8:245–249.

Flack, J. C., M. Girvan, F. B. M. de Waal, and D. C. Krakauer. 2006. Policing stabilizes construction of social niches in primates. *Nature* 439:426–429.

Flack, J. C., D. C. Krakauer, and F. B. M. de Waal. 2005. Robustness mechanisms in primate societies: a perturbation study. *Proc. R. Soc. Lond. B* 272:1091–1099.

Ford, J. K. B., G. M. Ellis, and K. C. Balcomb. 2000. *Killer whales*, 2nd ed. Vancouver, Canada: UBC Press.

Foster, B. R., and E. T. Rahs. 1983. Mountain goat response to hydroelectric exploration in northwestern British Columbia. *Environ. Manage.* 7:189–197.

Freeman, L. C. 1979. Centrality in social networks. Conceptual clarification. *Social Networks* 1:215–239.

Fresneau, D., and P. Dupuy. 1988. A study of polyethism in a ponerine ant: *Neoponera apicalis* (Hymenoptera, Formicidae). *Anim. Behav.* 36:1389–1399.

Frid, A., and L. M. Dill. 2002. Human-caused disturbance stimuli as a form of predation risk. *Conserv. Ecol.* 6:11.

Gammell, M. P., H. de Vries, D. J. Jennings, C. M. Carlin, and T. J. Hayden. 2003. David's score: a more appropriate dominance ranking method than Clutton-Brock et al.'s index. *Anim. Behav.* 66:601–605.

Gaskin, D. E. 1982. *The ecology of whales and dolphins*. London: William Heinemann.

Geissmann, T., and C. Braendle. 1997. Helping behaviour in captive pileated gibbons (*Hylobates pileatus*). *Folia Primatol.* 68:110–112.

Gero, S. 2005. Fundamentals of sperm whale societies: care for calves. MSc Thesis, Dalhousie University, Halifax, Nova Scotia.

Gero, S., and H. Whitehead. 2007. Suckling behavior in sperm whale calves: observations and hypotheses. *Mar. Mammal Sci.* 23:398-413.

Gero, S., L. Bejder, H. Whitehead, J. Mann, and R. C. Connor. 2005. Behaviourally specific preferred associations in bottlenose dolphins, *Tursiops* sp. *Can. J. Zool.* 83:1566–1573.

Ginsberg, J. R., and T. P. Young. 1992. Measuring association between individuals or groups in behavioural studies. *Anim. Behav.* 44:377–379.

Gompper, M. E., J. L. Gittleman, and R. K. Wayne. 1998. Dispersal, philopatry, and genetic relatedness in a social carnivore: comparing males and females. *Mol. Ecol.* 7:157–163.

Goodall, J. 1986. *The chimpanzees of Gombe: patterns of behavior*. Cambridge, MA: Harvard University Press.

———1968. Behaviour of free-living chimpanzees of the Gombe Stream Reserve. *Anim. Behav. Monogr.* 1:163–311.

Gowans, S., H. Whitehead, and S. K. Hooker. 2001. Social organization in northern bottlenose whales (*Hyperoodon ampullatus*): not driven by deep water foraging? *Anim. Behav.* 62:369–377.

Grafen, A. 1991. Modelling in behavioural ecology. In *Behavioural ecology. An evolutionary approach*, 3rd ed., ed. J. R. Krebs, and N. B. Davies. Oxford: Blackwell Scientific, pp. 5–31.

Hamilton, W. D. 1964. The genetical evolution of social behaviour. *J. Theor. Biol.* 7:1–52.

Hammond, P. S. 1986. Estimating the size of naturally marked whale populations using capture–recapture techniques. *Rep. Int. Whal. Commn. (Spec.
Iss.)* 8:253–282.

Harcourt, A. H., P. H. Harvey, S. G. Larson, and R. V. Short. 1981. Testis
weight, body weight and breeding system in primates. *Nature* 293:55–
57.

Hasselquist, D., and P. W. Sherman. 2001. Social mating systems and extrapair
fertilizations in passerine birds. *Behav. Ecol.* 12:457–466.

Hemelrijk, C. K. 1990a. A matrix partial correlation test used in investigations
of reciprocity and other social interaction patterns at group level. *J. Theor.
Biol.* 143:405–420.

———1990b. Models of, and tests for, reciprocity, unidirectionality and
other social interaction patterns at a group level. *Anim. Behav.* 39:1013–
1029.

Henzi, S. P., D. Lusseau, T. Weingrill, C. P. van Schaik, and L. Barrett. Submitted. Food availability influences female relationships within baboon troops.

Hill, R. A., and P. C. Lee. 1998. Predation risk as an influence on group size in
cercopithecoid primates: implications for social structure. *J. Zool.* (Lond.)
245:447–456.

Hinde, R. A. 1982. *Ethology. Its nature and relation with other sciences.* Oxford: Oxford University Press.

———1976. Interactions, relationships and social structure. *Man* 11:1–17.

Holme, P., S. M. Park, B. J. Kim, and C. R. Edling. 2007. Korean university life
in a network perspective: dynamics of a large affiliation network. *Physica A*
373:821–830.

Humphrey, N. K. 1976. The social function of intellect. In *Growing points in
ethology*, ed. P. P. G. Bateson, and R. A. Hinde. Cambridge: Cambridge
University Press, pp. 303–17.

International Whaling Commission. 1990. *Individual recognition of cetaceans:
use of photoidentification and other techniques to estimate population parameters. Reports of the International Whaling Commission Special Issue
12.* Cambridge: International Whaling Commission.

Isbell, L. A., and J. D. Pruetz. 1998. Differences between vervets (*Cercopithecus
aethiops*) and patas monkeys (*Erythrocebus patas*) in agonistic interactions
between adult females. *Int. J. Primatol.* 19:837–855.

Iverson, S. J., C. Field, W. D. Bowen, and W. Blanchard. 2004. Quantitative
fatty acid signature analysis: a new method of estimating predator diets.
Ecol. Monogr. 74:211–235.

Janik, V. M. 2001. Is social learning unique? *Behav. Brain Sci.* 24:337–338.

Jankowski, M. 2005. Long-finned pilot whale movement and social structure:
residency, population mixing and identification of social units. MSc Thesis,
Dalhousie University, Halifax, Nova Scotia.

Jarman, P. J. 1982. Prospects for interspecific comparisons in sociobiology. In
Current problems in sociobiology, ed. King's College Sociobiology Group.
Cambridge: Cambridge University Press. pp. 323–342.

———1974. The social organization of antelope in relation to their ecology.
Behaviour 48:215–267.

Johnson, D. H. 1999. The insignificance of statistical significance testing. *J. Wildl. Manage.* 63:763–772.

Kappeler, P. M., and C. P. van Schaik. 2002. Evolution of primate social systems. *Int. J. Primatol.* 23:707–740.

Karczmarski, L., B. Würsig, G. Gailey, K. W. Larson, and C. Vanderlip. 2005. Spinner dolphins in a remote Hawaiian atoll: social grouping and population structure. *Behav. Ecol.* 16:675–685.

Kinnaird, M. F., and T. G. O'Brien. 1996. Ecotourism in the Tangkoko DuaSudara nature reserve: opening Pandora's box? *Oryx* 30:65–73.

Kirk, R. E. 1995. *Experimental design: procedures for the behavioral sciences*, 3rd ed. Pacific Grove, CA: Brooks/Cole.

Kleinbaum, D. G., L. L. Kupper, and K. E. Muller. 1988. *Applied regression analysis and other multivariable methods*, 2nd ed. Boston: PWS-Kent.

Krause, J., D. P. Croft, and R. James. 2007. Social network theory in the behavioural sciences: potential applications. *Behav. Ecol. Sociobiol.* 62:15–27.

Krebs, C. J. 1989. *Ecological methodology.* New York: Harper and Row.

Krebs, J. R., and N. B. Davies. 1991. *An introduction to behavioural ecology*, 3rd ed. Oxford: Blackwell Scientific.

Krützen, M., J. Mann, M. R. Heithaus, R. C. Connor, L. Bejder, and W. B. Sherwin. 2005. Cultural transmission of tool use in bottlenose dolphins. *Proc. Natl. Acad. Sci. USA* 102:8939–8943.

Krützen, M., W. B. Sherwin, R. C. Connor, L. M. Barré, T. Van de Casteele, J. Mann, and R. Brooks. 2003. Contrasting relatedness patterns in bottlenose dolphins (*Tursiops* sp.) with different alliance strategies. *Proc. R. Soc. Lond. B* 270:497–502.

Kummer, H., W. Gotz, and W. Angst. 1974. Triadic differentiation: an inhibitory process protecting pair bonds in baboons. *Behaviour* 49:62–87.

Laland, K. N., and W. Hoppitt. 2003. Do animals have culture? *Evol. Anthropol.* 12:150–159.

Laland, K. N., and V. M. Janik. 2006. The animal cultures debate. *Trends Ecol. Evol.* 21:542–547.

Laland, K. N., J. Odling-Smee, and M. W. Feldman. 2000. Niche construction, biological evolution and cultural change. *Behav. Brain Sci.* 23:131–175.

Landau, H. G. 1951. On dominance relations and the structure of animal societies: I Effect of inherent characteristics. *Bull. Math. Biophys.* 13:1–19.

le Boeuf, B. J. 1974. Male–male competition and reproductive success in elephant seals. *Am. Zool.* 14:163–176.

Legendre, P., and L. Legendre. 1998. *Numerical ecology*, 2nd ed. Amsterdam: Elsevier.

Lehner, P. N. 1998. *Handbook of ethological methods*, 2nd ed. Cambridge: Cambridge University Press.

Leicht, E. A., P. Holme, and M. E. J. Newman. 2006. Vertex similarity in networks. *Phys. Rev. E* 73:026120.

Le Pendu, Y., L. Briedermann, J. F. Gerard, and M. L. Maublanc. 1995. Interindividual associations and social-structure of a mouflon population (*Ovis orientalis* Musimon). *Behav. Proc.* 34:67–80.

Li, M., Y. Fan, D. Wang, N. Liu, D. Li, J. Wu, and Z. Di. 2006. Effects of weight on structure and dynamics in complex networks. *arXiv:cond mat* 0601495.

Lindenfors, P., B. S. Tullberg, and M. Biuw. 2002. Phylogenetic analyses of sexual selection and sexual size dimorphism in pinnipeds. *Behav. Ecol. Sociobiol.* 52:188–193.

Lorenz, K. 1937. The companion in the bird's world. *Auk* 54:245–273.

———1935. Der Kumpan in der Umwelt des Vogels. *J. Ornithol.* 83:137–213, 289–413.

Lusseau, D. 2007. Why are male social relationships complex in the Doubtful Sound bottlenose dolphin population? *PLoS ONE* 2(4):e348.

———2003. The emergent properties of a dolphin social network. *Proc. R. Soc. Lond. B* 270:S186–S188.

Lusseau, D., and M. E. J. Newman. 2004. Identifying the role that animals play in social networks. *Proc. R. Soc. Lond. B* 271:S477–S481.

Lusseau, D., H. Whitehead, and S. Gero. In press. Applying network methods to the study of animal social structures. *Anim. Behav.*

Lusseau, D., K. Schneider, O. J. Boisseau, P. Haase, E. Slooten, and S. M. Dawson. 2003. The bottlenose dolphin community of Doubtful Sound features a large proportion of long-lasting associations. Can geographic isolation explain this trait? *Behav. Ecol. Sociobiol.* 54:396–405.

Manly, B. F. J. 1997. *Randomization, bootstrap and Monte Carlo methods in biology*, 2nd ed. London: Chapman and Hall.

———1995. A note on the analysis of species co-occurrences. *Ecology* 76:1109–1115.

———1994. *Multivariate statistical methods*, 2nd ed. New York: Chapman and Hall.

Mann, J. 2000. Unraveling the dynamics of social life: long-term studies and observational methods. In *Cetacean societies. Field studies of dolphins and whales*, ed. J. Mann, R. C. Connor, P. L. Tyack, and H. Whitehead. Chicago: University of Chicago Press, pp. 45–64.

———1999. Behavioral sampling methods for cetaceans: a review and critique. *Mar. Mammal Sci.* 15:102–122.

Mann, J., and B. Sargeant. 2003. Like mother, like calf: the ontogeny of foraging traditions in wild Indian ocean bottlenose dolphins (*Tursiops* sp.). In *The biology of traditions; models and evidence*, ed. D. M. Fragaszy, and S. Perry. Cambridge: Cambridge University Press, pp. 236–266.

Mann, J., and B. B. Smuts. 1998. Natal attraction: allomaternal care and mother–infant separations in wild bottlenose dolphins. *Anim. Behav.* 55:1097–1113.

Mantel, N. 1967. The detection of disease clustering and a generalized regression approach. *Cancer Res.* 27:209–220.

Marino, L. 1996. What can dolphins tell us about primate evolution? *Evol. Anthropol.* 5:81–85.

Martin, P., and P. Bateson. 2007. *Measuring behaviour: an introductory guide*, 3rd ed. Cambridge: Cambridge University Press.

Maryanski, A. R. 1987. African ape social structure: is there strength in weak ties? *Social Networks* 9:191–215.

McComb, K., C. Moss, S. M. Durant, L. Baker, and S. Sayialel. 2001. Matriarchs as repositories of social knowledge in African elephants. *Science* 292:491–494.

McConnell, B., M. Fedak, J. Matthiopoulos, and P. Lovell. 2003. Telemetry: bits, models, phones and dust. Abstract. In *15th Biennial Conference on the Biology of Marine Mammals, Greensboro, North Carolina, December 2003*, p. 107.

McCormick, M. I., and S. Smith. 2004. Efficacy of passive integrated transponder tags to determine spawning-site visitations by a tropical fish. *Coral Reefs* 23:570–577.

McGrew, W. C. 1992. *Chimpanzee material culture: implications for human evolution.* Cambridge: Cambridge University Press.

McMahan, C. A., and M. D. Morris. 1984. Application of maximum likelihood paired comparison ranking to estimation of a linear dominance hierarchy in animal societies. *Anim. Behav.* 32:374–378.

Michener, C. D. 1969. Comparative social behavior of bees. *Ann. Rev. Entomol.* 144:299–342.

Michener, G. R. 1980. The measurement and interpretation of interaction rates: an example with adult Richardson's ground squirrels. *Biol. Behav.* 5:371–384.

Milligan, G. W., and M. C. Cooper. 1987. Methodology review: clustering methods. *Appl. Psychol. Meas.* 11:329–354.

Mitani, J. C., G. F. Grether, P. S. Rodman, and D. Priatna. 1991. Associations among wild orang-utans: Sociality, passive aggregations or chance? *Anim. Behav.* 42:33–46.

Moehlman, P. D. 1998. Feral asses (*Equus africanus*): intraspecific variation in social organization in arid and mesic habitats. *Appl. Anim. Behav. Sci.* 60:171–195.

Mohtashemi, M., and L. Mui. 2003. Evolution of indirect reciprocity by social information: the role of trust and reputation in evolution of altruism. *J. Theor. Biol.* 223:523–531.

Morgan, B. J. T., M. J. A. Simpson, J. P. Hanby, and J. Hall-Craggs. 1976. Visualizing interaction and sequential data in animal behaviour: theory and application of cluster-analysis methods. *Behaviour* 56:1–43.

Mundinger, P. C. 1980. Animal cultures and a general theory of cultural evolution. *Ethol. Sociobiol.* 1:183–223.

Murchison, C. 1935. The experimental measurement of a social hierarchy in *Gallus domesticus*: IV. Loss of body weight under conditions of mild starvation as a function of social dominance. *J. Gen. Psychol.* 12:296–312.

Myers, J. P. 1983. Space, time and the pattern of individual associations in a group-living species: sanderlings have no friends. *Behav. Ecol. Sociobiol.* 12:129–134.

Nagel, U. 1979. On describing primate groups as systems. In *Primate ecology and human origins: ecological influences on social organization*, ed. I. S. Bernstein, and E. O. Smith. New York: Garland STPM Press, pp. 313–339.

Newman, M. E. J. 2006. Modularity and community structure in networks. *Proc. Natl. Acad. Sci. USA* 103:8577–8582.

———2004. Analysis of weighted networks. *Phys. Rev. E* 70:056131.

————2003a. Mixing patterns in networks. *Phys. Rev. E* 67:026126.

————2003b. The structure and function of complex networks. *SIAM Rev.* 45:167–256.

Newman, M. E. J., and E. A. Leicht. 2006. Mixture models and exploratory data analysis in networks. arXiv:physics/0611158v2.

Noldus Information Technology. 2003. *MatMan, reference manual, version 1.1.* Wageningen, Netherlands: Noldus.

Ottensmeyer, C. A., and H. Whitehead. 2003. Behavioural evidence for social units in long-finned pilot whales. *Can. J. Zool.* 81:1327–1338.

Packer, C., D. A. Gilbert, A. E. Pusey, and S. J. O'Brien. 1991. A molecular genetic analysis of kinship and co-operation in African lions. *Nature* 351:562–565.

Packer, C., D. Scheel, and A. E. Pusey. 1990. Why lions form groups: food is not enough. *Am. Nat.* 136:1–19.

Palla, G., A.-L. Barabási, and T. Vicsek. 2007. Quantifying social group evolution. *Nature* 446:664–667.

Palla, G., I. Derényi, I. Farkas, and T. Vicsek. 2005. Uncovering the overlapping community structure of complex networks in nature and society. *Nature* 435:814–818.

Palsbøll, P. J., J. Allen, M. Bérubé, P. J. Clapham, T. P. Feddersen, P. S. Hammond, R. R. Hudson, H. Jørgensen, S. Katona, A. H. Larsen, F. Larsen, J. Lien, D. K. Matilla, J. Sigurjónsson, R. Sears, T. Smith, R. Sponer, P. Stevick, and N. Φien. 1997. Genetic tagging of humpback whales. *Nature* 388:767.

Payne, K. 2003. Sources of social complexity in the three elephant species. In *Animal social complexity; intelligence, culture, and individualized societies,* ed. F. B. M. de Waal and P. L. Tyack. Cambridge, MA: Harvard University Press, pp. 57–85.

————1999. The progressively changing songs of humpback whales: a window on the creative process in a wild animal. In *The origins of music,* ed. N. L. Wallin, B. Merker, and S. Brown. Cambridge, MA: MIT Press, pp. 135–50.

Pennycuick, C. J. 1978. Identification using natural markings. In *Animal marking. Recognition marking of animals in research,* ed. B. Stonehouse. Baltimore: University Park Press, pp. 147–159.

Pepper, J. W., J. C. Mitani, and D. P. Watts. 1999. General gregariousness and specific social preferences among wild chimpanzees. *Int. J. Primatol.* 20:613–632.

Perry, S. 1996. Female–female social relationships in wild white-faced capuchin monkeys, *Cebus capucinus. Am. J. Primatol.* 40:167–182.

Perry, S., and J. H. Manson. 2003. Traditions in monkeys. *Evol. Anthropol.* 12:71–81.

Pfeiffer, T., C. Rutte, T. Killingback, M. Taborsky, and S. Bonhoeffer. 2005. Evolution of cooperation by generalized reciprocity. *Proc. R. Soc. Lond. B* 272:1115–1120.

Pitcher, T. J. 1986. Functions of shoaling behaviour in teleosts. In *The Behaviour of teleost fishes,* ed. T. J. Pitcher. London: Cro.om Helm, pp. 294–337.

Poole, J. H., and J. Thomsen. 1989. Elephants are not beetles: implications of the ivory trade for the survival of the African elephant. *Oryx* 23:188–198.

Proulx, S. R., D. E. L. Promislow, and P. C. Phillips. 2005. Network thinking in ecology and evolution. *Trends Ecol. Evol.* 6:345–353.

Pulliam, H. R., and T. Caraco. 1984. Living in groups: is there an optimal group size? In *Behavioural ecology. An evolutionary approach*, 2nd ed., ed. J. R. Krebs, and N. B. Davies. Oxford: Blackwell Scientific, pp. 122–47.

Purvis, A., and A. Rambaut. 1995. Comparative analysis by independent contrasts (CAIC): an Apple Macintosh application for analysing comparative data. *CABIOS* 11:247–251.

Queller, D. C., and K. F. Goodnight. 1989. Estimating relatedness using genetic markers. *Evolution* 43:258–275.

Randerson, P. F. 1993. Ordination. In *Biological data analysis*, ed. J. C. Fry. Oxford: Oxford. University Press, pp. 173–217.

Reader, S. M., and L. Lefebvre. 2001. Social learning and sociality. *Behav. Brain Sci.* 40:353–354.

Rendell, L. E., and H. Whitehead. 2003. Comparing repertoires of sperm whale codas: a multiple methods approach. *Bioacoustics* 14:61–81.

Rendell, L., and H. Whitehead. 2001. Cetacean culture: still afloat after the first naval engagement of the culture wars. *Behav. Brain Sci.* 24:360–373.

Richerson, P. J., and R. Boyd. 2004. *Not by genes alone: how culture transformed human evolution.* Chicago: University of Chicago Press.

Richerson, P. J., and R. Boyd. 1998. The evolution of human ultrasociality. In *Indoctrinability, ideology and warfare*, ed. I. Eibl-Eibesfeldt and. F. K. Salter. London: Berghahn Books, pp. 71–95.

Rogers, L. M., R. Delahay, C. L. Cheeseman, S. Langton, G. C. Smith, and R. S. Clifton-Hadley. 1998. Movement of badgers (*Meles meles*) in a high-density population: individual, population and disease effects. *Proc. R. Soc. Lond. B* 265:1269–1276.

Roney, J. R., and D. Maestripieri. 2003. Social development and affiliation. In *Primate psychology*, ed. D. Maestripieri. Cambridge, MA: Harvard University Press, pp. 171–204.

Roper, T. J. 1986. Cultural evolution of feeding behaviour in animals. *Sci. Prog.* 70:571–583.

Rossiter, S. J., G. Jones, R. D. Ransome, and E. M. Barratt. 2002. Relatedness structure and kin-based foraging in the greater horseshoe bat (*Rhinolophus ferrumequinum*). *Behav. Ecol. Sociobiol.* 51:510–518.

Rowell, T. E. 1979. How would we know if social organization were not adaptive? In *Primate ecology and human origins*, ed. I. S. Bernstein, and E. O. Smith. New York: Garland/STPM Press, pp. 1–22.

———1972. *Social behaviour of monkeys.* London: Penguin Books.

Ruxton, G. D., and N. Colegrave. 2006. *Experimental design for the life sciences*, 2nd ed. Oxford: Oxford University Press.

Sattath, S., and A. Tversky. 1977. Additive similarity trees. *Psychometrika* 42:319–345.

Schnell, G. D., D. J. Watt, and M. E. Douglas. 1985. Statistical comparison of proximity matrices: applications in animal behaviour. *Anim. Behav.* 33:239–253.

Schusterman, R. J., C. J. Reichmuth, and D. Kastak. 2000. How animals classify friends and foes. *Curr. Direct. Psychol. Sci.* 9:1–6.

Seber, G. A. F. 1992. A review of estimating animal abundance II. *Int. Stat. Rev.* 60:129–166.

———1982. *The estimation of animal abundance and related parameters*, 2nd ed. London: Griffin.

Seghers, B. H. 1974. Schooling behaviour in the guppy (*Poecilia reticulata*): an evolutionary response to predation. *Evolution* 28:486–489.

Selkoe, K. A., and R. J. Toonen. 2006. Microsatellites for ecologists: a practical guide to using and evaluating microsatellite markers. *Ecol. Lett.* 9:615–629.

Shannon, C. E., and W. Weaver. 1949. *The mathematical theory of communication.* Urbana, IL: University of Illinois Press.

Sibly, R. M. 1983. Optimal group size is unstable. *Anim. Behav.* 31:947–948.

Silvey, S. D. 1975. *Statistical inference.* London: Chapman and Hall.

Singh, M., and L. D'Souza. 1992. Hierarchy, kinship and social interaction among Japanese monkeys (*Macaca fuscata*). *Ind. J. Dermatol.* 37:15–27.

Slater, P. 1961. Inconsistencies in a schedule of paired comparisons. *Biometrika* 48:303–312.

Sloane, M. A., P. Sunnucks, D. Alpers, L. B. Beheregaray, and A. C. Taylor. 2000. Highly reliable genetic identification of individual hairy-nosed wombats from single remotely collected hairs: a feasible censusing method. *Mol. Ecol.* 9:1233–1240.

Smolker, R. A., A. F. Richards, R. C. Connor, and J. W. Pepper. 1992. Sex differences in patterns of association among Indian Ocean bottlenose dolphins. *Behaviour* 123:38–69.

Smouse, P. E., J. C. Long, and R. R. Sokal. 1986. Multiple regression and correlation extensions of the Mantel test of matrix correspondence. *Syst. Zool.* 35:627–632.

Smuts, B. B. 1985. *Sex and friendship in baboons.* New York: Aldine.

Sokal, R. R., and F. J. Rohlf. 1994. *Biometry*, 3rd ed. New York: W. H. Freeman.

Soltis, J., R. Boyd, and P. J. Richerson. 1995. Can group-functional behaviors evolve by cultural group selection? An empirical test. *Curr. Anthropol.* 36:473–494.

Stander, P. E. 1992. Cooperative hunting in lions: the role of the individual. *Behav. Ecol. Sociobiol.* 29:445–454.

Starck, J.-L., and F. Murtagh. 1999. Multiscale entropy filtering. *Signal Proc.* 76:147–165.

Stearns, S. C., and R. F. Hoekstra. 2000. *Evolution.* Oxford: Oxford University Press.

Stonehouse, B. 1978. *Animal marking. Recognition marking of animals in research.* Baltimore: University Park Press.

Strauss, R. E. 2001. Cluster analysis and the identification of aggregations. *Anim. Behav.* 61:481–488.

Sugiyama, Y. 1988. Grooming interactions among adult chimpanzees at Bossou, Guinea, with special reference to social structure. *Int. J. Primatol.* 9:393–407.

Sumpter, D. J. T. 2006. The principles of collective animal behaviour. *Philos. Trans. R. Soc. Lond. (B Biol. Sci.)* 361:5–22.

Sunobe, T. 2000. Social structure, nest guarding and interspecific relationships of the cichlid fish (*Julidochromis marlieri*) in Lake Tanganyika. *Afr. Study Monogr.* 21:83–89.

Sutherland, W. J. 1998. The importance of behavioural studies in conservation biology. *Anim. Behav.* 56:801–809.

Thompson, W. R. 1958. Social behavior. In *Behavior and evolution*, ed. A. Roe and G. G. Simpson. New Haven, CT: Yale University Press, pp. 291–310.

Tietjen, G. L. 1986. *A topical dictionary of statistics*. New York: Chapman and Hall.

Tinbergen, N. 1963. On aims and methods of ethology. *Z. Tierpsychol.* 20:410–433.

———1953. *Social behaviour in animals, with special reference to vertebrates.* London: Methuen.

Trivers, R. 1985. *Social evolution.* Menlo Park, CA: Benjamin/Cummings.

———1971. The evolution of reciprocal altruism. *Q. Rev. Biol.* 46:35–57.

Underwood, R. 1981. Companion preference in an eland herd. *Afr. J. Ecol.* 19:341–354.

van de Casteele, T., P. Galbusera, and E. Mattysen. 2001. A comparison of microsatellite-based pairwise relatedness estimators. *Mol. Ecol.* 10:1539–1549.

van Hooff, J. A. R. A. M., and J. A. B. Wensing. 1987. Dominance and its behavioral measures in a captive wolf pack. In *Man and wolf*, ed. H. Frank. Dordrecht, Holland: Junk, pp. 219–252.

van Schaik, C. P., and D. M. Kappeler. 1997. Infanticide risk and the evolution of male–female association in primates. *Proc. R. Soc. Lond. B* 264:1687–1694.

van Schaik, C. P., M. Ancrenaz, G. Borgen, B. Galdikas, C. D. Knott, I. Singleton, A. Suzuki, S. S. Utami, and M. Merrill. 2003. Orangutan cultures and the evolution of material culture. *Science* 299:102–105.

Vervaecke, H., H. de Vries, and L. van Elsacker. 2000. The pivotal role of rank in grooming and support behavior in a captive group of bonobos (*Pan paniscus*). *Behaviour* 137:1463–1485.

Voegeli, F. A., M. J. Smale, D. M. Webber, Y. Andrade, and R. K. O'Dor. 2001. Ultrasonic telemetry, tracking and automated monitoring technology for sharks. *Environ. Biol. Fish.* 60:267–282.

Vonhof, M. J., H. Whitehead, and M. B. Fenton. 2004. Analysis of Spix's disk-winged bat association patterns and roosting home ranges reveal a novel social structure among bats. *Anim. Behav.* 68:507–521.

Wedderburn, R. W. M. 1974. Quasi-likelihood functions, generalized linear models, and the Gauss–Newton method. *Biometrika* 61:439–447.

Weinrich, M. T., H. Rosenbaum, C. S. Baker, A. L. Blackmer, and H. Whitehead. 2006. The influence of maternal lineages on social affiliations among humpback whales (*Megaptera novaeangliae*) on their feeding grounds in the Southern Gulf of Maine. *J. Hered.* 97:226–234.

Wellman, B., and S. D. Berkowitz. 2003. *Social structures: a network approach.*
Toronto: Canadian Scholars' Press.

Wey, T., D. T. Blumstein, W. Shen, and F. Jordán. 2008. Social network analysis of animal behaviour: a promising tool for the study of sociality. *Anim. Behav.*

Whitehead, H. In press-a. Precision and power in the analysis of social structure using associations. *Anim. Behav.*

———In press-b. How might we study culture? A perspective from the ocean. In *The question of animal culture*, ed. R. N. Laland and B. Galef. Cambridge, MA: Harvard University Press.

———2007. Selection of models of lagged identification rates and lagged association rates using AIC and QAIC. *Commun. Statist. Simul. Comput.* 36:1233–1246.

———2004. The group strikes back: follow protocols for behavioral research on cetaceans. *Mar. Mammal Sci.* 20:304–310.

———2003. *Sperm whales: social evolution in the ocean.* Chicago: University of Chicago Press.

———2001. Analysis of animal movement using opportunistic individual-identifications: application to sperm whales. *Ecology* 82:1417–1432.

———1999a. Testing association patterns of social animals. *Anim. Behav.* 57:F26–F29.

———1999b. Variation in the visually observable behavior of groups of Galápagos sperm whales. *Mar. Mammal Sci.* 15:1181–1197.

———1997. Analyzing animal social structure. *Anim. Behav.* 53:1053–1067.

———1995. Investigating structure and temporal scale in social organizations using identified individuals. *Behav. Ecol.* 6:199–208.

———1983. Structure and stability of humpback whale groups off Newfoundland. *Can. J. Zool.* 61:1391–1397.

Whitehead, H., and T. Arnbom. 1987. Social organization of sperm whales off the Galápagos Islands, February–April 1985. *Can. J. Zool.* 65:913–919.

Whitehead, H., and S. Dufault. 1999. Techniques for analyzing vertebrate social structure using identified individuals: review and recommendations. *Adv. Study Behav.* 28:33–74.

Whitehead, H., and T. Wimmer. 2005. Heterogeneity and the mark-recapture assessment of the Scotian Shelf population of northern bottlenose whales (*Hyperoodon ampullatus*). *Can. J. Fish. Aquat. Sci.* 62:2573–2585.

Whitehead, H., L. Bejder, and A. C. Ottensmeyer. 2005. Testing association patterns: issues arising and extensions. *Anim. Behav.* 69:e1–e6.

Whitehead, H., M. Dillon, S. Dufault, L. Weilgart, and J. Wright. 1998. Non–geographically based population structure of South Pacific sperm whales: dialects, fluke-markings and genetics. *J. Anim. Ecol.* 67:253–262.

Whitehead, H., S. Waters, and T. Lyrholm. 1991. Social organization in female sperm whales and their offspring: constant companions and casual acquaintances. *Behav. Ecol. Sociobiol.* 29:385–389.

Whitehead, H., R. Silver, and P. Harcourt. 1982. The migration of humpback whales along the northeast coast of Newfoundland. *Can. J. Zool.* 60:2173–2179.

Whiten, A. 2000. Social complexity and social intelligence. In *The nature of intelligence*. Chichester, U.K.: Wiley, pp. 185–201.

Whiten, A., and R. Ham. 1992. On the nature and evolution of imitation in the animal kingdom: reappraisal of a century of research. *Adv. Study Behav.* 21:239–283.

Whiten, A., J. Goodall, W. C. McGrew, T. Nishida, V. Reynolds, Y. Sugiyama, C. E. G. Tutin, R. W. Wrangham, and C. Boesch. 2001. Charting cultural variation in chimpanzees. *Behaviour* 138:1481–1516.

Whiten, A., J. Goodall, W. C. McGrew, T. Nishida, V. Reynolds, Y. Sugiyama, C. E. G. Tutin, R. W. Wrangham, and C. Boesch. 1999. Cultures in chimpanzees. *Nature* 399:682–685.

Whitham, J. C., and D. Maestripieri. 2003. Primate rituals: the function of greetings between male guinea baboons. *Ethology* 109:847–859.

Williams, C. K., R. S. Lutz, and R. D. Applegate. 2003. Optimal group size and northern bobwhite coveys. *Anim. Behav.* 66:377–387.

Wilson, E. B. 1927. Probable inference, the law of succession, and statistical inference. *J. Am. Stat. Assoc.* 22:209–212.

Wilson, E. O. 1975. *Sociobiology: the new synthesis*. Cambridge, MA: Belknap Press.

——— 1971. *The insect societies*. Cambridge, MA: Belknap Press.

Wittemyer, G., I. Douglas-Hamilton, and W. M. Getz. 2005. The socio-ecology of elephants: analysis of the processes creating multi-tiered social structures. *Anim. Behav.* 69:1357–1371.

Wrangham, R. W. 1980. An ecological model of female-bonded primate groups. *Behaviour* 75:262–300.

Wrangham, R. W., and D. I. Rubenstein. 1986. *Ecological aspects of social evolution*. Princeton, NJ: Princeton University Press.

Wrangham, R. W., A. P. Clark, and G. Isabirye-Basuta. 1992. Female social relationships and social organization of Kibale Forest chimpanzees. In *Topics in primatology*, ed. T. Nishida, W. C. McGrew, P. Marler, M. Pickford, and F. B. M. de Waal. Tokyo: University of Tokyo Press, pp. 81–98.

Yurk, H., L. Barrett-Lennard, J. K. B. Ford, and C. O. Matkin. 2002. Cultural transmission within maternal lineages: vocal clans in resident killer whales in southern Alaska. *Anim. Behav.* 63:1103–1119.

Zahavi, A. 1977. The testing of a bond. *Anim. Behav.* 25:246–247.

Zeller, D. C. 1999. Ultrasonic telemetry: its application to coral reef fisheries research. *Fish. Bull. US* 97:1058–1065.

Zumpe, D., and R. P. Michael. 1986. Dominance index: a simple measure of relative dominance status in primates. *Int. J. Primatol.* 10:291–300.

Index

classification, statistical methods of.
 See cluster analysis
classifying; relationships, 16–17, 110–
 12, 131, 221; social structures, 5,
 8, 15, 17–18, 242–44
cluster analysis, 41–44, 151–54, 161–
 65, 167–68, 224, 238, 239, 252,
 253, 304, 305; delineating communities, groups, tiers, and units,
 226–32, 236–37
clustering coefficient, 148, 174–75,
 176, 179–81, 183, 296
Clutton-Brock et al. dominance index,
 191, 194
coalitions, 135, 136, 275. *See also*
 alliances
coefficient of variation (CV), 26, 29
coherence. *See* half-weight association
 index
collecting social data, 64–71
collinearity in regression analysis, 287
communication, 6, 18–19, 56, 59,
 264–65
communities; closure of, 223, 244;
 comparisons of, 20, 64, 267–
 69; definition of, 14–15, 222,
 296; delineating, 221, 222–40,
 305; defining classes of animals,
 64; elements of cluster analysis,
 41; as replicators, 274; size of, as
 measure of social complexity, 18,
 146, 247, 248, 271–73; size of,
 estimating, 223, 306–8
compartmentalization. *See* modularity
competition for resources, 256–58
complete-linkage cluster analysis, 43,
 165, 227
complexity, social, 9, 18, 20, 219,
 246–50
computer programs, 46–51, 292, 293,
 301–5
conceptual frameworks of social analysis, 11–14, 15–16, 289–90
conflict, 256–58
conformism, 277, 288
conservation, 4
cooperation, 4, 119, 258–65

cophenetic correlation coefficient
 (CCC), 44, 162, 163, 165, 167,
 237, 304
copulations, 257, 258, 276–77, 278
correlation coefficient (r), 26, 29–30,
 267, 268
correlation, partial, 122
correlogram, 107
correspondence analysis, 26, 38–39,
 41, 252, 253, 254
covariance, 26, 29
Crow's method for dominance ranks,
 189, 193, 194
culture, 13, 14, 246, 277, 279–88;
 definition, 277, 296; determining
 cultural behavior, 280–87; conformist, 277, 289; group selection
 and, 274, 277; identifying cultural
 entities, 279–80
cyclical patterns, 105, 203, 206

database software, 46–49, 71, 79.
 See also Access, Microsoft; data
 formats
data collection, 53–79
data formats, 71–79; dyadic mode,
 72–73, 74, 75, 76, 296; group
 mode, 73, 74, 75, 76, 77, 297;
 linear mode, 72, 73, 74, 76, 77,
 297; supplemental data, 77–79
data matrices, 35–38
data sets; attributes of, 80–81, 290–
 92; size of, 22, 23, 81–86, 290,
 291, 292
data storage software, 47, 48–49. *See
 also* data formats
David's dominance index, 188, 190–
 91, 194, 195
deep structure, 11, 14. *See also* social
 structure
defense of resources, 271, 275
degree, network measure, 172, 176,
 177, 296
demography, 17, 18, 247, 249; affecting association indices, 99,
 103, 104, 147; affecting lagged
 association rates, 203; controlling